Undergraduate Texts in Mathematics

Reading in Mathematics

Editors

S. Axler
F.W. Gehring
K.A. Ribet

Springer

New York
Berlin
Heidelberg
Barcelona
Hong Kong
London
Milan
Paris
Singapore
Tokyo

Graduate Texts in Mathematics
Readings in Mathematics

Undergraduate Texts in Mathematics
Readings in Mathematics

John Stillwell

Numbers and Geometry

With 95 Illustrations

 Springer

John Stillwell
Monash University
Department of Mathematics
Clayton, Victoria
31 68 Australia

Cover illustration of the exponential map by Tristan Needham. Reprinted from *Visual Complex Analysis* by permission of Tristan Needham and Oxford University Press, 1997.

Mathematics Subject Classification (2000): 11-01, 51-01

Library of Congress Cataloging-in-Publication Data
Stillwell, John.
    Numbers and geometry
        p.   cm. — (Undergraduate texts in mathematics. Readings in
    mathematics)
        Includes bibliographical references (p.   –   ) and index.
        ISBN 0-387-98289-2 (hc; alk. paper)
        1. Mathematics.   I. Title.   II. Series.
    QA39.2.S755   1997
    510—dc21
                                        97-22858
                                           CIP

Printed on acid-free paper.

Production managed by Lesley Poliner; manufacturing supervised by Joe Quatela.
Photocomposed copy prepared from the author's TEX files.
Printed and bound by Maple-Vail Book Manufacturing Group, York, PA.
Printed in the United States of America.

9 8 7 6 5 4 3 2

ISBN 0-387-98289-2 Springer-Verlag New York Berlin Heidelberg   SPIN 10764915

To my mother and my late father
who taught me how to count

Preface

What should every aspiring mathematician know? The answer for most of the 20th century has been: calculus. For 2000 years before that, the answer was Euclid. It now seems a good time to raise the question again, because the old answers are no longer convincing. Mathematics today is much more than Euclid, but it is also much more than calculus; and the calculus now taught is, sadly, much less than it used to be. Little by little, calculus has been deprived of the algebra, geometry, and logic it needs to sustain it, until many institutions have had to put it on high-tech life-support systems. A subject struggling to survive is hardly a good introduction to the vigor of real mathematics.

But if it were only a matter of putting the guts back into calculus it would not be necessary to write a new book. It would be enough to recommend, for example, Spivak's *Calculus*, or Hardy's *Pure Mathematics*. In the current situation, we need to revive not only calculus, but also algebra, geometry, and the whole idea that mathematics is a rigorous, cumulative discipline in which each mathematician stands on the shoulders of giants.

The best way to teach real mathematics, I believe, is to start deeper down, with the elementary ideas of number and space. Everyone concedes that these are fundamental, but they have been scandalously neglected, perhaps in the naive belief that anyone learning calculus has outgrown them. In fact, arithmetic, algebra, and geometry can never be outgrown, and the most rewarding path to higher mathematics sustains their development alongside the "advanced" branches such as calculus. Also, by maintaining ties between these disciplines, it is possible to present a more unified view of mathematics, yet at the same time to include more spice and variety.

The aim of this book, then, is to give a broad view of arithmetic, geometry, and algebra at the level of calculus, without being a calculus book (or a "precalculus" book). Its roots are in arithmetic and geometry, the two opposite poles of mathematics, and the source of historic conceptual conflict. The resolution of this conflict, and its role in the development of mathematics, is one of the main stories in the book. The key is algebra, which brings arithmetic and geometry together and allows them to flourish and branch out in new directions. To keep the story as simple as possible, I link everything to the algebraic themes of linear and quadratic equations.

The restriction to low-degree equations is not as dreadful as high school algebra might suggest. Even linear equations are interesting when only integer solutions are sought, and they neatly motivate a whole introductory course in number theory, from the Euclidean algorithm to unique prime factorization. Quadratic equations are even more interesting from the integer point of view, with Pythagorean triples and Pell's equation leading deep into algebra, geometry, and analysis. From the point of view of geometry, quadratic equations represent the conic sections—a fascinating topic in itself—and the areas bounded by these curves define the circular, logarithmic, and hyperbolic functions. In this way we are led to the subject matter of a first calculus course, but with less machinery and more time to probe fundamental questions such as the nature of numbers, curves, and area. It is worth mentioning that we also cover the main ideas of Euclid—geometry, arithmetic, and the theory of real numbers—but with 2000 years of extra insights added.

In fact, this book could be described as a deeper look at ordinary things. Most of mathematics is about numbers, curves, and functions, and the links between these concepts can be suggested by a thorough study of simple examples, such as the circle and the square. I hope to show that mathematics, like the world, fits William Blake's description:

> To see a World in a Grain of Sand,
> And a Heaven in a Wild Flower,
> Hold Infinity in the palm of your hand
> And Eternity in an hour.

Because it is virtually impossible to learn mathematics by mere reading, this book includes many exercises and every encouragement to do them. There is a set of exercises at the end of each section, so new ideas can be instantly tested, clarified, and reinforced. The exercises are often variations or generalizations of the results in the main text; in some cases I think they are even more interesting! In particular, they include simplified arrangements of many classic proofs by great mathematicians, from Euclid to Hilbert. Each set of exercises is accompanied by a commentary to make its purpose and significance clear. I hope this will be useful, particularly when several exercises have to be linked together to produce a big result.

Who is this book for? Because it presupposes only high school algebra, it can in principle be read by any well-prepared student entering university. It complements the usual courses, hence it can be offered as a "hard option" to students who are not sufficiently extended by the standard material at that level. On the other hand, it has so little in common with the calculus and linear algebra that dominates the standard curriculum that it may well be a revelation even to senior undergraduates. Many students now graduate in mathematics without having done a course in number theory, geometry, or foundations—for example, without having seen the fundamental theorem of arithmetic, non-Euclidean geometry, or the definition of real numbers. For such students, this book could serve as a capstone course in the senior year, presenting a unified approach to mathematics and proving many of the classic results that are normally only mentioned.

It could also be used in conventional courses. Chapters 1, 4, 6, 7, and parts of 8 and 9 contain most of the standard material for a first number theory course. Chapters 2, 3, 5, and 8 could serve as a first course in geometry. But naturally it would be best if the two courses were coordinated—perhaps run in parallel—to take advantage of the links between them. The whole is greater than the sum of its parts.

A glance at the table of contents and the index will reveal that this book contains a lot of material, some of it quite hard. This is because I want to provide many interesting paths to follow, for students of all levels. However, it is not necessary to follow each path to its end. The harder sections and exercises are marked with stars, and they can be omitted without losing access to most of the material that follows. There are also informal discussions at the ends of chapters, intended to help readers see the big picture even while some of the details remain confused. Each discussion deals with a few main themes, placing them in historical and mathematical perspective, linking them to other parts of the book, and sometimes extending their development and suggesting further reading.

The book grew out of a talk I gave at Oberwolfach in November 1995, following a suggestion of Urs Kirchgraber. Several parts of it have been used in courses at Monash, from first year to fourth year, and the book in its entirety has benefited from the comments of Mark Aarons, Benno Artmann, Tristan Needham, and Aldo Taranto. To them, and as usual to my wife Elaine, I offer my sincere thanks.

Clayton, Victoria, Australia   John Stillwell

February 1997

Contents

1

Arithmetic

1.1 The Natural Numbers

The beauty and fascination of numbers can be summed up by one simple fact: anyone can count $1, 2, 3, 4, \ldots$, but no one knows all the implications of this simple process. Let me elaborate. We all realize that the sequence $1, 2, 3, 4, \ldots$ continues $5, 6, 7, 8, \ldots$, and that we can continue *indefinitely* adding 1. The objects produced by the counting process are what mathematicians call the *natural numbers*. Thus if we want to say what it is that $1, 2, 3, 17, 643, 100097801$, and 4514517888888856 have in common, in short, what a natural number *is*, we can only say that each is produced by the counting process. This is slightly troubling when you think about it: the simplest, and most finite, mathematical objects are defined by an infinite process. However, the concept of *natural number* is inseparable from the concept of infinity, so we must learn to live with it and, if possible, use it to our advantage.

In fact, one of the most powerful methods in mathematics draws its strength from the infinite counting process. This is *mathematical induction*, which we usually just call *induction* for short. It may be formulated in several ways, each basically a restatement of the fact that any natural number can be reached by counting.

1

The first form of induction we consider (and apparently the first actually used) expresses the fact that from each natural number we can "count down" to 1, by finitely often subtracting 1. It follows that an infinite descending sequence of natural numbers is *impossible*. And *nonexistence* of natural numbers with certain properties often follows by hypothetical construction of an infinite descending sequence. This form of induction is called *infinite descent*,[1] or simply *descent*. Possibly the oldest example is the following, which goes back to around 500 B.C. To abbreviate the proof and show its simple logical structure, we use the symbol $\Rightarrow$ for "hence" or "implies."

The proof shows that no natural number square is twice another, but the result is better known as the "irrationality of $\sqrt{2}$."

Irrationality of $\sqrt{2}$ *There are no natural numbers m and n such that* $m^2 = 2n^2$.

Proof The hypothetical equation $m^2 = 2n^2$ leads to a similar equation, but with smaller numbers, as follows:

$$m^2 = 2n^2 \Rightarrow m^2 \text{ even}$$
$$\Rightarrow m \text{ even, say, } m = 2m_1$$
$$\Rightarrow 4m_1^2 = m^2 = 2n^2$$
$$\Rightarrow n^2 \text{ even}$$
$$\Rightarrow n \text{ even, say, } n = 2n_1$$
$$\Rightarrow m_1^2 = 2n_1^2 \text{ and } m > m_1 > 0$$
$$\Rightarrow m_2^2 = 2n_2^2 \text{ and } m > m_1 > m_2 > 0, \text{ similarly,}$$

and so on. Thus we get an infinite descending sequence, $m > m_1 > m_2 > \cdots$, which is impossible. Hence there are no natural numbers m and n such that $m^2 = 2n^2$. □

As most readers will know, $\sqrt{2}$ is defined as the number x such that $2 = x^2$. The proof shows that $\sqrt{2}$ is *not a ratio m/n* of natural numbers m, n, as this would imply $2 = m^2/n^2$ and hence $m^2 = 2n^2$. This is why $\sqrt{2}$ is called *irrational*; it simply means "not a ratio." As is typical when we wish to prove a negative statement, we argue by

---

[1]The unsettling experience of infinite descent has been used as the basis of a horror story by Marghanita Laski, called *The Tower*.

contradiction: the existence of a ratio $m/n = \sqrt{2}$ is shown to imply an impossibility.

But if $\sqrt{2}$ is not a ratio, what is it? Does it even exist? These questions have had an enormous influence on the development of mathematics, and their answers fill a large part of this book. For the moment, it is enough to say that, whatever the whole story of $\sqrt{2}$ may be, its irrationality is a fact about natural numbers.

Exercises

A problem at least as old as the meaning of $\sqrt{2}$, though less subtle, also leads to an interesting descent argument. Again, this is a case where a question about fractions reduces to a question about natural numbers.

About 4000 years ago, the Egyptians invented a curious arithmetic of fractions that depended on expressing each fraction between 0 and 1 as a sum of distinct *unit fractions*, that is, fractions of the form $\frac{1}{n}$. For example, $\frac{2}{3}$ is the sum of the unit fractions $\frac{1}{2}$ and $\frac{1}{6}$. Such sums are called *Egyptian fractions*. As another example, an Egyptian fraction for $\frac{3}{5}$ is $\frac{1}{2} + \frac{1}{10}$.

1.1.1. Express $\frac{4}{5}$, $\frac{9}{10}$, and $\frac{11}{12}$ as Egyptian fractions.

1.1.2. Find two different Egyptian fractions for $\frac{7}{12}$.

We do not know the Egyptian methods for finding such sums. They seem to involve many special tricks for avoiding unnecessarily large denominators, and it is difficult to capture them all in a rule that works in all cases. A more systematic approach was developed in the book *Liber Abacci*, written in 1202 by Leonardo of Pisa, better known as Fibonacci.

The method of the *Liber Abacci* also includes several tricks, but one of them can be used on its own to express any fraction between 0 and 1 as an Egyptian fraction. The trick is to repeatedly *remove the largest unit fraction*. Thus if $\frac{a}{b}$ is a (nonunit) fraction between 0 and 1, in lowest terms, let $\frac{1}{n}$ be the largest unit fraction less than $\frac{a}{b}$, and form the new fraction $\frac{a'}{b'} = \frac{a}{b} - \frac{1}{n}$.

1.1.3. Assuming $\frac{a'}{b'}$ is in lowest terms, show that $0 < a' < a$.

1.1.4. Hence conclude, by descent, that finitely many such removals split $\frac{a}{b}$ into distinct unit fractions.

1.1.5. Use repeated removal of the largest unit fraction to show

$$\frac{6}{19} = \frac{1}{4} + \frac{1}{16} + \frac{1}{304}.$$

It is worth mentioning that Fibonacci obtained the simpler decomposition $\frac{6}{19} = \frac{1}{4} + \frac{1}{19} + \frac{1}{76}$. After removal of the largest unit fraction, $\frac{1}{4}$, he was left with $\frac{5}{76}$. Rather than repeat the process of removing the largest unit fraction, he took advantage of the fact that 76 has the divisor 4 to split $\frac{5}{76}$ into $\frac{4}{76} + \frac{1}{76} = \frac{1}{19} + \frac{1}{76}$.

It is also worth mentioning that there are still some unsolved problems with Egyptian fractions. For example, it is not known whether each fraction of the form $\frac{4}{n}$ is the sum of three or fewer unit fractions. (For more information on problems with Egyptian fractions, see Guy (1994).)

1.2   Division, Divisors, and Primes

So far we have taken addition, multiplication, and fractions more or less for granted, and we shall continue to do so until a deeper investigation is called for (Section 1.9*). However, we cannot take division for granted, because it cannot always be done in the natural numbers. As you learned in primary school, 3 into 7 "won't go," so we are forced to consider the more complicated concept of *division with remainder*. The exact relation between 3 and 7 is that

$$7 = 2 \times 3 + 1,$$

which we express by saying that 2 is the *quotient* when 7 is divided by 3, and 1 is the *remainder*. Only when there is no remainder, as when 3 divides 6, is true division possible in the natural numbers.

If a and b are any natural numbers, we say that b *divides* a if there is a natural number q such that

$$a = qb.$$

In this case, we also say that a is *divisible* by b, or that b is a *divisor* of a, or that a is a *multiple* of b.

If b does not divide a then $a \neq b, 2b, 3b, \ldots$, hence if we descend through the numbers $a, a - b, a - 2b, a - 3b, \ldots$ we eventually reach (because we cannot descend indefinitely), a natural number $r = a - qb$ smaller than b. We then have the result

$$a = qb + r, \quad \text{with } r < b.$$

The natural number q is called the quotient on division of a by b, and r is called the remainder. The fact that a can be expressed as a multiple of b plus a remainder smaller than b is often called the *division algorithm*, though we prefer to use that name for the *process* of division (namely, repeated subtraction of b until the remainder is smaller than b), and call the fact the *division property*.

The relation of division with remainder includes true division, of course, when we allow $r = 0$. In fact, some people include 0 among the natural numbers, but it is helpful to distinguish it as a new number: the first of several extensions of the number concept.[2] The fractions, for example, are an extension of the natural numbers because the natural numbers n are just the special fractions $\frac{n}{1}$. At this point you may wonder why we do not move to fractions immediately and make division easier. After all,

$$7 = \frac{7}{3} \times 3,$$

so 3 *does* divide 7 in the world of fractions. The reason is that fractions do not overcome the difficulty of division, they only conceal it. The problem comes back when we have to decide when a fraction is in lowest terms. We know $\frac{6}{3}$ is not in lowest terms, for example, because we know that 3 divides 6 *in the natural numbers*.

This example helps to clarify what is "natural" about the natural numbers. Apart from being the medium for counting, they are also the natural setting for division, divisors, and *factorization* —the process of writing numbers as products. When a natural number is written as a product, say,

$$a = n_1 n_2 \cdots n_k,$$

---

[2] The only disadvantage in taking 0 to be new is that there is then no name for the enlarged set "natural numbers together with 0." This is only a temporary inconvenience; we soon need further extensions of the natural numbers, which *do* have names.

the divisors n_j of a are called *factors*. The simplest numbers, from the standpoint of factorization, are the *primes* —the natural numbers p divisible only by 1 and p. They may be regarded as "atoms" because they cannot be split into smaller factors. Factors of 1 are redundant, so 1 is not classed as a prime. The first few prime numbers are

$$2, 3, 5, 7, 11, 13, 17, 19, 23, 29, 31, 37, 41, 43, 47, 53, 59, 61, \dots .$$

Exercises

In the proof that $\sqrt{2}$ is irrational (Section 1.1), we used the fact that m^2 is even if and only if m is even, or in other words that 2 divides m^2 if and only if 2 divides m. This is easily checked, but it is worth spelling out, because algebra is involved, and the idea of "algebraic factorization" has many other applications.

$$2 \text{ divides } m \Rightarrow m = 2l \text{ for some } l$$
$$\Rightarrow m^2 = (2l)^2 = 2(2l^2)$$
$$\Rightarrow 2 \text{ divides } m^2.$$

$$2 \text{ does not divide } m \Rightarrow m = 2l + 1 \text{ for some } l$$
$$\Rightarrow m^2 = (2l + 1)^2 = 4l^2 + 4l + 1 = 2(2l^2 + 2l) + 1$$
$$\Rightarrow 2 \text{ does not divide } m^2.$$

This idea has a generalization to multiples of 3.

1.2.1. Show that m^2 is a multiple of 3 only if m is a multiple of 3. Hence prove that there are no natural numbers m and n such that $m^2 = 3n^2$.

This proves the irrationality of $\sqrt{3}$; there are other ways to prove it, some of which are more general and apply to $\sqrt{5}, \sqrt{6}, \dots$ as well. We shall see them in Section 1.6. Another important algebraic factorization is the following.

1.2.2. Check that $x^n - 1 = (x - 1)(x^{n-1} + x^{n-2} + \cdots + x + 1)$.

This enables us to find divisors of certain large numbers.

1.2.3. Deduce from Exercise 1.2.2 that, if $m = np$, then $2^m - 1$ is divisible by $2^p - 1$.

1.2.4. Conclude that $2^p - 1$ is prime only if p is prime.

Primes of the form $2^p - 1$ are known as *Mersenne primes* after Marin Mersenne (1588-1648), who first drew attention to them. About 35 Mersenne primes have been found, but it is not known whether there are infinitely many.

1.2.5. Check that $2^p - 1$ is prime when $p = 2, 3, 5, 7$, but not when $p = 11$.

1.2.6.* By similarly factorizing $x^n + 1$ when n is odd, deduce that $2^m + 1$ is prime only if m has no odd divisors, that is, only if m is a power of 2.

1.2.7. Check that $2^{2^h} + 1$ is prime for $h = 0, 1, 2, 3, 4$, but that 641 divides $2^{2^5} + 1$.

Primes of the form $2^{2^h} + 1$ are called *Fermat primes*, after Pierre de Fermat. Apart from those with $h = 0, 1, 2, 3, 4$, no other Fermat primes are known.

1.3   The Mysterious Sequence of Primes

It is relatively easy to continue the list of primes, especially with the help of a computer, but one never gets a clear picture of where it is going. Somehow, the two simplest aspects of the natural numbers—their ability to be ordered and their ability to be factored—interact in an incredibly complex way. Listing the primes in increasing order produces no apparent pattern; one cannot even be sure the list continues indefinitely. On the other hand, the *concept* of prime is surely simple, so maybe we can prove that there are infinitely many primes, without knowing their pattern.

This is in fact what Euclid did, more than 2000 years ago. You can read his simple proof in Proposition 20, Book IX of the *Elements*, which is available in English in the excellent edition of Heath (1925). Here is a slightly modernized version.

Euclid's Theorem   *There are infinitely many primes.*

Proof   First we need to see that any natural number n has a prime divisor.

Take any divisor d of n. If d is prime, we have found a prime divisor. If not, d has a smaller divisor $e \neq 1$. This divisor e of d is also a divisor of n, because $n = dq$ and $d = er$ for some natural numbers q and r, and therefore $n = erq$. If e is not prime, we repeat the argument, finding a smaller divisor $f \neq 1$. Because we cannot descend indefinitely, we eventually find a prime divisor of n.

Now we use a prime divisor to extend any given list of primes. Given primes $p_1, p_2, \ldots, p_k$, consider the number

$$n = p_1 p_2 \cdots p_k + 1.$$

This number is not divisible by any of the given primes $p_1, p_2, \ldots, p_k$, because they all leave remainder 1. But we have just seen that n has some prime divisor p. Thus, if $p_1, p_2, \ldots, p_k$ are any given primes, we can find a prime $p \neq p_1, p_2, \ldots, p_k$. □

This proof is one of the most admired in mathematics, and one's admiration for it only increases the more one knows about primes. Euclid, like us, did not know any pattern in the sequence of primes, so he devised a proof that did not *need* to know. If he and later mathematicians had waited for someone to find a pattern, we still would not know the first thing about primes.

Exercises

Euclid's proof is the simplest way to see that infinitely many primes *exist*, though not the most practical way to find them. Still, it is fun to produce new primes by multiplying known primes together and adding 1. Starting with the single prime $p_1 = 2$, for example, we get $n = 2 + 1 = 3$, which is a second prime p_2. Then p_1, p_2 give $n = 2 \times 3 + 1 = 7$, which is a third prime p_3; p_1, p_2, p_3 give $n = 2 \times 3 \times 7 + 1 = 43$, and so on.

1.3.1. Continue this process, and find the first stage where $n = p_1 p_2 p_3 \cdots p_k + 1$ is not itself a prime.

If you take the *least* prime divisor of n at each stage, you should be able to continue long enough to find an n in this sequence whose least prime divisor is 5. With some computer help, you might be able to continue long enough to reach an n whose least prime divisor is 11. (It is preceded by some huge prime values of n.) It is not known whether *each* prime is eventually produced by this process.

Apart from the number 2, all prime numbers are odd, and odd numbers are of two types: those of the form $4n + 1$ and those of the form $4n + 3$. It turns out to be helpful to split the odd primes in this way as well, because the two types of odd prime often behave differently. For a start, we can extend Euclid's idea to prove that there are infinitely many primes of the form $4n + 3$.

1.3.2. Show that the product of $4a + 1$ and $4b + 1$ is a number of the form $4n + 1$.

1.3.3. Deduce from Exercise 1.3.2 that any number of the form $4m + 3$ has a prime factor of the form $4n + 3$.

1.3.4. Show that $p_1, p_2, \ldots, p_k$ do not divide $2p_1 p_2 \cdots p_k + 1$.

1.3.5. Show, however, that if $p_1, p_2, \ldots, p_k$ are all odd primes then *some prime of the form* $4n + 3$ divides $2p_1 p_2 \cdots p_k + 1$. Deduce that there are infinitely many primes of the form $4n + 3$.

It is also true that there are infinitely many primes of the form $4n + 1$, but this is harder to prove. The best possible result in this direction was proved by Peter Lejeune Dirichlet (1837). He showed that any sequence of the form $an + b$, where a and b are natural numbers with no common divisor, contains infinitely many primes. For example, Dirichlet's theorem says there are infinitely many primes of the form $6n + 1$ and of the form $6n + 5$, but there are none of the form $6n + 3$ (because 3 divides any number of the form $6n + 3$). In general, if a and b have a common divisor, there are no primes of the form $an + b$.

The form $an + b$ is called a *linear* form, so Dirichlet's theorem settles the question of how many primes there are in a given linear form. Virtually nothing is known about primes in higher-degree forms. For example, we do not know whether there are infinitely many primes of the form $n^2 + 1$.

1.4 Integers and Rationals

Everyone will agree that the natural numbers $1, 2, 3, 4, \ldots$ deserve the name "natural," but mathematicians feel they are not natural enough. $1, 2, 3, 4, \ldots$ are fine for counting, but not for arithmetic, because they do not permit unlimited subtraction. We cannot take

7 from 3, for example. To make this possible, we extend the set $\mathbb{N}$ of natural numbers to the set $\mathbb{Z}$ of *integers*[3] by adjoining 0 and the *negative integers* $-1, -2, -3, -4, \ldots$. The negative integers can be viewed as the result of subtracting $1, 2, 3, 4, \ldots$ respectively from 0, but it is simpler to regard attachment of the negative sign as the basic operation, and to *define* subtraction by $a - b = a + (-b)$.

The natural numbers now start a new life as the *positive integers*. Each positive integer a has an *additive inverse* $-a$, and the additive inverse of $-a$ is defined to be a. If we also define $-0 = 0$, then it follows that in all cases $-(-a) = a$.

The integers are a more natural home for arithmetic because they permit addition, subtraction, and multiplication without restriction. However, questions arise about the meaning of these operations on the newly introduced numbers. What is $(-1) - (-4)$ for example, or $(-1) \times (-1)$? The best way to answer these questions is by "keeping things natural." We ask ourselves how $+$, $-$, and $\times$ behave on $\mathbb{N}$ and insist that they behave the same on $\mathbb{Z}$.

First, we can summarize how $+$ and $-$ behave by the following rules, which hold for all positive integers a, b and c:

$$a + (b + c) = (a + b) + c \qquad \text{(associative law)}$$
$$a + b = b + a \qquad \text{(commutative law)}$$
$$a + (-a) = 0 \qquad \text{(additive inverse property)}$$
$$a + 0 = a \qquad \text{(identity property of 0)}$$

These are nothing but the rules we use unconsciously when doing addition and subtraction on positive integers. We have to become conscious of them now, to see what they imply for integers in general.

It follows, for example, that we have *uniqueness of additive inverse*: $-a$ is the *only* integer b such that $a + b = 0$. This is what we normally call "solving for b," but with more awareness of the individual steps:

$$a + b = 0 \Rightarrow (-a) + (a + b) = -a \qquad \text{adding } -a \text{ to both sides}$$
$$\Rightarrow ((-a) + a) + b = -a \qquad \text{by the associative law}$$
$$\Rightarrow (a + (-a)) + b = -a \qquad \text{by the commutative law}$$

---

[3]The letter $\mathbb{Z}$ is the initial of the German word "Zahlen" meaning "numbers."

$\Rightarrow 0 + b = -a$ by the additive inverse property

$\Rightarrow b = -a$ by the identity property of 0.

There is a similar set of rules describing the behavior of $\times$:

$$a \times (b \times c) = (a \times b) \times c \qquad \text{(associative law)}$$
$$a \times b = b \times a \qquad \text{(commutative law)}$$
$$a \times 1 = a \qquad \text{(identity property of 1)}$$
$$a \times 0 = 0 \qquad \text{(property of 0)}$$

and finally, a rule for the interaction of $+$ and $\times$:

$$a \times (b + c) = a \times b + a \times c \qquad \text{(distributive law)}$$

From these we deduce that $a \times (-1) = -a$ for any integer a because

$$a + a \times (-1) = a \times 1 + a \times (-1) \qquad \text{by the identity property of 1}$$
$$= a \times (1 + (-1)) \qquad \text{by the distributive law}$$
$$= a \times 0 \qquad \text{by the additive inverse property}$$
$$= 0 \qquad \text{by the property of 0}$$
$$\Rightarrow \quad a \times (-1) = -a \qquad \text{by the uniqueness of additive inverse.}$$

It follows in particular that $(-1) \times (-1) = 1$, because $-(-1) = 1$.

We extend the set $\mathbb{Z}$ of integers to the set $\mathbb{Q}$ of *rational numbers,*[4] or simply *rationals*, by adjoining a *multiplicative inverse* a^{-1} of each nonzero integer a. The multiplicative inverse of a^{-1} is defined to be a, and these inverses have the following property:

$$a \times a^{-1} = 1 \qquad \text{(multiplicative inverse property)}$$

These properties of $\mathbb{Q}$ are what we use unconsciously in doing ordinary arithmetic with $+$, $-$, $\times$, and $\div$. The quotient $a \div b$ or a/b is the same as $a \times b^{-1}$. As mentioned earlier, questions about the arithmetic of $\mathbb{Q}$ are really equivalent to questions about $\mathbb{Z}$, or even $\mathbb{N}$, but the extra properties of $\mathbb{Q}$ sometimes make life easier. This is particularly the case in geometry, where the rational numbers pave the way for interpreting *points* as numbers.

---

[4]The symbol $\mathbb{Q}$ stands for "quotients." We do not use the initial letter of "rational" because the same letter is later needed for the real numbers.

Exercises

The rules governing the behavior of $+$, $-$, and $\times$ are called the *ring properties* of $\mathbb{Z}$, and in general any set with functions $+$ and $\times$ satisfying these rules is called a *ring*. As we have already said, the ring properties of $\mathbb{Z}$ are so familiar that we normally use them unconsciously. Becoming conscious of them helps us to understand arithmetic, not only in $\mathbb{Z}$, but also in any other system that satisfies the same rules. We call such a system a *commutative ring with unit*, the "unit" in this case being the number 1. Later we shall find it helpful to use many other rings, even to study $\mathbb{Z}$ itself.

The following exercises help to explain why the ring properties are fundamental to arithmetic, by showing how they determine the values of expressions written using natural numbers, $+$, $-$, and $\times$, and some standard algebraic identities.

1.4.1. Show, using the properties of $+$ and $-$, that $(-1) - (-4) = 3$.

1.4.2. More generally, show that $(-a) - (-b) = b - a$.

1.4.3. Now, using properties of $\times$ and the distributive law, show $(-a)(-b) = ab$.

1.4.4. Also use the ring properties to prove that

$$(a + b)^2 = a^2 + 2ab + b^2 \qquad \text{and} \qquad (a - b)(a + b) = a^2 - b^2,$$

where, as usual, xy stands for $x \times y$.

Incidentally, there is a good mathematical reason for abbreviating $x \times y$ to xy. The distributive law is better written as

$$a(b + c) = ab + ac,$$

because the products on the right-hand side have precedence over the sum—they have to be evaluated first.

A ring with the multiplicative inverse property, such as $\mathbb{Q}$, is called a *field*. In fact, the way we extended $\mathbb{Z}$ to $\mathbb{Q}$ is an instance of a common construction with rings, called "forming the field of fractions." For any a and $b \neq 0$ we form the fraction $a/b = ab^{-1}$, and we add and multiply fractions according to the rules you learned around fifth grade:

$$\frac{a}{b} + \frac{c}{d} = \frac{ad + bc}{bd} \qquad \text{and} \qquad \frac{a}{b}\frac{c}{d} = \frac{ac}{bd}$$

These rules also arise from the principle of "keeping things natural"—they are needed to make + and × behave the same for fractions as they do for natural numbers.

1.5 Linear Equations

The humble linear equation $ax + by = c$ takes on a new interest when we seek *integer* solutions x and y for given integers a, b, and c. It can very easily fail to have an integer solution, so the problem is first to decide *whether* there is an integer solution, and if so, how to find it.

Take the example $15x + 12y = 4$. For any integers x and y, 3 divides $15x + 12y$, because 3 divides 15 and 12. But 3 does not divide 4, hence there is no integer solution of $15x + 12y = 4$. In general, we can see that $ax + by = c$ has no integer solution if a and b have a divisor that does not divide c.

But what if the divisors of a and b divide c? It is not at all clear there are integers x and y with $ax + by = c$, though they seem to exist in every case we try. For example, if we consider $17x + 12y = 1$, the only divisors of both 17 and 12 are ± 1, which certainly divide the right-hand side. And with some difficulty (say, by searching down lists of the multiples of 17 and 12) we indeed find a solution, $x = 5$ and $y = -7$.

This presumably depends on some connection between divisors of a and b and numbers of the form $ax + by$. We can already see part of it: Any divisor of a and b is also a divisor of $ax + by$, for any integers x and y. The less obvious part comes from thinking about the *greatest* common divisor of a and b, which we call $\gcd(a, b)$, and seeing that it has the form $ax + by$.

There is a famous algorithm for finding $\gcd(a, b)$, for natural numbers a and b. It is called the *Euclidean algorithm*, and it was described by Euclid as "repeatedly subtracting the smaller number from the larger." To be precise, we produce pairs of natural numbers (a_1, b_1), (a_2, b_2), (a_3, b_3), ... as follows. The first pair (a_1, b_1) is (a, b) itself, and each new pair (a_{i+1}, b_{i+1}) comes from (a_i, b_i) by

$$a_{i+1} = \max(a_i, b_i) - \min(a_i, b_i) \qquad \text{(taking the smaller from the larger),}$$

$$b_{i+1} = \min(a_i, b_i) \qquad \text{(and keeping the smaller),}$$

until $a_k = b_k$, in which case the algorithm halts. Then $\gcd(a, b) = a_k = b_k$.

It is clear that the algorithm does reach a pair of equal numbers a_k and b_k, because the natural numbers $a_1, a_2, a_3, \ldots$ cannot decrease indefinitely. But why does it produce the gcd?

Correctness of the Euclidean algorithm   *All pairs produced by the Euclidean algorithm have the same common divisors, hence $a_k = b_k = \gcd(a, b)$.*

Proof   Each divisor of a_i and b_i is also a divisor of a_{i+1} (because any divisor of two numbers also divides their difference) and of b_{i+1}. Conversely, any divisor of a_{i+1} and b_{i+1} also divides a_i and b_i, because any divisor of two numbers also divides their sum. Thus each pair (a_{i+1}, b_{i+1}) has the same divisors as all previous pairs, and hence the same gcd. But then

$$\gcd(a, b) = \gcd(a_1, b_1) = \gcd(a_2, b_2) = \cdots = \gcd(a_k, b_k) = a_k = b_k,$$

because $a_k = b_k$.   □

Not only does the Euclidean algorithm give $\gcd(a, b)$, it gives it in the form $ax + by$.

Linear representation of the gcd   *All the numbers a_i, b_i produced by the Euclidean algorithm are of the form $ax + by$, for some integers x and y, hence this is also the form of $\gcd(a, b)$ itself.*

Proof   The first pair a, b are certainly each of the required form. This is also true of all subsequent numbers a_{i+1}, b_{i+1}, because each is either a previous number or the difference of two of them. In particular, $\gcd(a, b) = a_k = ax + by$ for some integers x and y.   □

We illustrate the Euclidean algorithm on $a = 17$ and $b = 12$ in the first two columns of the following table. The third column keeps track of what happens to a and b, eventually giving x and y with $17x + 12y = \gcd(17, 12) = 1$.

$$(a_1, b_1) = (17, 12) = (a, b)$$
$$(a_2, b_2) = (5, 12) = (a - b, b)$$
$$(a_3, b_3) = (7, 5) = (b - (a - b), a - b)$$
$$= (2b - a, a - b)$$
$$(a_4, b_4) = (2, 5) = ((2b - a) - (a - b), a - b)$$
$$= (3b - 2a, a - b)$$
$$(a_5, b_5) = (3, 2) = ((a - b) - (3b - 2a), 3b - 2a)$$
$$= (3a - 4b, 3b - 2a)$$
$$(a_6, b_6) = (1, 2) = ((3a - 4b) - (3b - 2a), 3b - 2a)$$
$$= (5a - 7b, 3b - 2a)$$

The last line shows the gcd, 1, expressed as $5a - 7b$. Thus we have the solution $x = 5$, $y = -7$ to $17x + 12y = 1$, just as we found by trial before. The Euclidean algorithm does not have any computational advantage in a small example such as this, but it does in large examples. If a and b are integers with many digits, the Euclidean algorithm can be completed in roughly as many steps as there are digits (see the exercises), whereas listing the multiples of a and b takes an exponentially larger number of steps.

In addition to being computationally powerful, the Euclidean algorithm gives us remarkable theoretical insight. For a start, we have confirmed our guess about integer solutions $ax + by = c$.

Test for integer solvability of $ax + by = c$ *The equation $ax + by = c$ has an integer solution if and only if $\gcd(a, b)$ divides c.*

Proof   We have already seen that if $\gcd(a, b)$ does *not* divide c, then the equation $ax + by = c$ has no integer solution.

Conversely, if $\gcd(a, b)$ divides c, suppose $c = \gcd(a, b) \times d$. We now know that there are integers x' and y' such that $\gcd(a, b) = ax' + by'$. Therefore, $c = (ax' + by')d = a(x'd) + b(y'd)$, and hence we have the solution $x = x'd$, $y = y'd$ of $ax + by = c$. $\square$

Exercises

In practice, we usually speed up the Euclidean algorithm by dividing the larger number by the smaller and keeping the remainder, instead of subtracting the smaller number from the larger. (Halting then occurs when the remainder is 0.) Because division of a by b is really subtraction of b from a until the difference is less than b, the division form of the algorithm produces the same result—it simply skips any steps where the same number is subtracted more than once. This saves many steps when a is much larger than b.

1.5.1. Show that the remainder, when a is divided by a smaller number b, is less than $a/2$.

1.5.2. Deduce from Exercise 1.5.1 that the number of steps to find $\gcd(a, b)$, by the division form of the Euclidean algorithm, is at most twice the number of binary digits in a.

An interesting showcase for the Euclidean algorithm is the *Fibonacci sequence*, $0, 1, 1, 2, 3, 5, 8, 13, 21, 34, 55, \ldots$, in which each number is the sum of the previous two.

1.5.3. Use the Euclidean algorithm to verify that $\gcd(55, 34) = 1$.

We use the notation F_n for the nth term of the Fibonacci sequence, starting with $F_0 = 0$. The whole sequence can then be defined by the equations

$$F_0 = 0, \quad F_1 = 1, \quad F_{n+1} = F_n + F_{n-1}.$$

1.5.4. Show that one step of the Euclidean algorithm on the pair (F_{n+2}, F_{n+1}) produces the pair (F_{n+1}, F_n), hence

$$1 = \gcd(F_2, F_1) = \gcd(F_3, F_2) = \gcd(F_4, F_3) = \cdots.$$

You probably also noticed that "division of F_{n+2} by F_{n+1}" is really subtraction, so in the case of consecutive Fibonacci numbers, the subtraction form of the Euclidean algorithm cannot be sped up. In fact, the Euclidean algorithm performs at its slowest on consecutive Fibonacci numbers, though it would take us too far afield to explain what this means. The full story may be found in Shallit (1994).

Because $\gcd(F_{n+1}, F_n) = 1$ by Exercise 1.5.4, it follows by the corollary to the correctness of the Euclidean algorithm that there are integers x and y such that $F_{n+1}x + F_ny = 1$.

1.5.5. Find integers x_1 and y_1 such that $F_2x_1 + F_1y_1 = 1$, and integers x_2 and y_2 such that $F_3x_2 + F_2y_2 = 1$.

1.5.6. Show that $F_{n+2}F_n - F_{n+1}F_{n+1} = -F_{n+1}F_{n-1} + F_nF_n$ and hence that

$$1 = -F_2F_0 + F_1F_1 = F_3F_1 - F_2F_2 = -F_4F_2 + F_3F_3 = \cdots.$$

It is worth mentioning that when Euclid proved that there are infinitely many primes, he did not argue exactly as we did in Section 1.3. Instead of using division with remainder to prove that each prime p_j fails to divide $p_1 p_1 \cdots p_k + 1$, he used the obvious fact that $\gcd(p_1 p_1 \cdots p_k + 1, p_1 p_1 \cdots p_k) = 1$.

1.5.7. Why is this obvious? Use the similar fact $\gcd(p_1 p_1 \cdots p_k - 1, p_1 p_1 \cdots p_k) = 1$ to give another proof that there are infinitely many primes.

1.6   Unique Prime Factorization

The discovery that the greatest common divisor of a and b is of the form $ax + by$, for some integers x and y, has important repercussions for prime divisors.

Prime divisor property   *If a prime p divides the product of integers a and b, then p divides either a or b.*

Proof   Suppose that p divides ab and p does *not* divide a. Then we have to show that p divides b. Because p does not divide a, and p is prime, 1 is the only divisor of p that divides a. We therefore have

$$1 = \gcd(a, p) = ax + py \qquad \text{for some integers } x \text{ and } y.$$

It follows, multiplying both sides by b, that

$$b = abx + pby.$$

But p divides each term on the right-hand side of this equation—it divides ab by assumption and pby obviously—hence p divides b.   □

   This important property was known to Euclid, as were many of its important consequences, which we shall see later. However, he did not state the following consequence, which today is considered the definitive statement about prime divisors.

Unique prime factorization   *Each natural number is expressible in only one way as a product of primes, apart from the order of factors.*

Proof By repeatedly finding prime divisors (Section 1.3), we can factorize any natural number into primes. Now suppose, contrary to the theorem, that there is a natural number with two different prime factorizations:

$$p_1 p_2 p_3 \cdots p_s = q_1 q_2 q_3 \cdots q_t.$$

We may assume that any factor common to both sides has already been canceled, hence no factor on the left is on the right.

But p_1 divides the left-hand side, and therefore it divides the right, which is a product of q_1 and $q_2 q_3 \cdots q_t$. Thus it follows from the prime divisor property that p_1 divides q_1 (in which case $p_1 = q_1$, because q_1 is prime) or else p_1 divides $q_2 q_3 \cdots q_t$. In the latter case we similarly find either $p_1 = q_2$ or p_1 divides $q_3 \cdots q_t$. Continuing in this way, we eventually find that

$$p_1 = q_1 \quad \text{or} \quad p_2 = q_2 \quad \text{or} \quad \cdots \quad \text{or} \quad p_1 = q_t.$$

But this contradicts our assumption that p_1 is not a prime on the right side. Thus there is no natural number with two different prime factorizations. □

A variation on the preceding proof, which some people prefer, starts with $p_1 p_2 p_3 \cdots p_s = q_1 q_2 q_3 \cdots q_t$ but does *not* assume that the factorizations are different. One again finds $p_1 = q_1$ or $p_2 = q_2$ or $\cdots$ or $p_1 = q_t$, but now this simply means that there is a common factor p_1 on both sides. Cancel it, and repeat until no primes remain, at which stage it is clear that the original factorizations were the same.

Exercises

Unique prime factorization is a powerful way to prove results like the irrationality of $\sqrt{2}$, which we first did in Section 1.1 using properties of even and odd numbers, that is, by using special properties of the number 2. We saw that to extend the method to $\sqrt{3}$ requires a new (and longer) argument about the number 3, and presumably it gets worse for $\sqrt{5}$, $\sqrt{6}$, and so on. With unique prime factorization, the argument depends only on the presence of primes, not which particular ones they are.

For example, to prove irrationality of $\sqrt{2}$, we observe that the equation

$$m^2 = 2n^2$$

contradicts unique prime factorization. Why? The prime 2 necessarily occurs an even number of times in the prime factorization of the left-hand side, namely, twice the number of times it occurs in m. But it occurs an odd number of times on the right-hand side: the visible occurrence, plus twice the number of times it occurs in n.

Exactly the same argument applies to the equation $m^2 = 3n^2$, but with the prime 3 in place of the prime 2, and hence proves the irrationality of $\sqrt{3}$. Likewise for the equation $m^2 = 5n^2$, and the irrationality of $\sqrt{5}$. The irrationality of $\sqrt{6}$ is a little different, of course, because 6 is not a prime. But in this case it still works to consider the prime factors in the hypothetical equation $m^2 = 6n^2$.

1.6.1. Prove the irrationality of $\sqrt{6}$, that is, the impossibility of $m^2 = 6n^2$.

The irrationality of many other numbers can be proved by the same idea—showing that a hypothetical equation has some prime occurring to different exponents on the left- and right-hand sides.

1.6.2. Prove the irrationality of $\sqrt[3]{2}$, that is, the impossibility of $m^3 = 2n^3$.

1.6.3. Prove the irrationality of $\log_{10} 2$, that is, the impossibility of $2 = 10^{m/n}$.

In the *Disquisitiones Arithmeticae* (arithmetical investigations) of Carl Friedrich Gauss (1801) there is an interesting direct proof of the prime divisor property, by descent.

1.6.4. First show that a prime p cannot divide a product of smaller numbers. Suppose that p divides $a_1 b_1$, where $a_1, b_1 < p$, and deduce that p also divides $a_1 b_2$, where

$$b_2 = \text{remainder when } p \text{ is divided by } b_1,$$

which gives an infinite descent.

1.6.5. Use Exercise 1.6.4 to deduce the prime divisor property, by showing that if p divides ab, and p divides neither a nor b, then p divides an $a_1 b_1$ with $a_1, b_1 < p$.

Gauss remarked that the prime divisor property was already proved by Euclid, however we did not wish to omit it, because many modern authors have offered up feeble arguments in place of proof or have neglected the theorem completely. *(Gauss (1801), article 14)*

1.7   Prime Factorization and Divisors

Unique prime factorization is called the *fundamental theorem of arithmetic*, and was first stated by Gauss (1801). Gauss also pointed out how unique prime factorization allows us to describe all the divisors of a given natural number.

For example, because $30 = 2 \times 3 \times 5$, the numbers $1, 2, 3, 5, 2 \times 3$, $2 \times 5, 3 \times 5$, and $2 \times 3 \times 5$ are all divisors of 30. Conversely, any natural number divisor a of 30 satisfies

$$2 \times 3 \times 5 = ab$$

for some natural number b. By uniqueness, the prime factorization of ab is also $2 \times 3 \times 5$, and a is part of it, hence a is one of the numbers listed.

In general, if

$$n = p_1^{e_1} p_2^{e_2} \cdots p_k^{e_k},$$

where $p_1, p_2, \ldots, p_k$ are the distinct prime divisors of n, and $e_1, e_2, \ldots, e_k$ are their exponents, then the natural number divisors of n are numbers of the form

$$n = p_1^{d_1} p_2^{d_2} \cdots p_k^{d_k},$$

where $0 \leq d_1 \leq e_1, 0 \leq d_2 \leq e_2, \ldots, 0 \leq d_k \leq e_k$. This is because the prime factorization of a divisor is (by uniqueness) part of the prime factorization of n.

It may be that general statements about prime factorization and divisors were not made by Euclid because he lacked a notation for exponents. The same goes for the following description of greatest common divisors and least common multiples, which first appear in Gauss (1801), although they were probably known much earlier. They follow immediately from the description of divisors in terms of prime factors. The idea of finding the $\gcd(m, n)$ by collecting all the common prime factors m and n is certainly an obvious one, sometimes taught in primary school, because it works well for small numbers.

$$\gcd(p_1^{e_1} p_2^{e_2} \cdots p_k^{e_k}, p_1^{f_1} p_2^{f_2} \cdots p_k^{f_k}) = p_1^{\min(e_1, f_1)} p_2^{\min(e_2, f_2)} \cdots p_k^{\min(e_k, f_k)}.$$

The least common multiple of m and n is abbreviated $\mathrm{lcm}(m, n)$, and we have

$$\mathrm{lcm}(p_1^{e_1} p_2^{e_2} \cdots p_k^{e_k}, p_1^{f_1} p_2^{f_2} \cdots p_k^{f_k}) = p_1^{\max(e_1, f_1)} p_2^{\max(e_2, f_2)} \cdots p_k^{\max(e_k, f_k)}.$$

Putting these two formulas together, we get the elegant formula

$$\gcd(m, n)\mathrm{lcm}(m, n) = mn,$$

which apparently was not noticed by Euclid. This formula shows, incidentally, how to compute $\mathrm{lcm}(m, n)$ *without* prime factorizations of m and n: compute $\gcd(m, n)$ by the Euclidean algorithm, then divide it into mn.

The climax of Euclid's number theory occurs at the end of Book IX of the *Elements*, where he proves a famous theorem about perfect numbers. A natural number n is called *perfect* if it is the sum of its proper divisors, that is, the natural number divisors apart from itself. The Greeks thought of the proper divisors as the "parts" of a number, hence a perfect number was the "sum of its parts." Only a few examples were then known, the smallest being $6 = 1 + 2 + 3$ and the next being $28 = 1 + 2 + 4 + 7 + 14$. Euclid found a general formula that includes these and all other known examples by finding all the divisors of numbers of the form $2^{n-1}p$, where p is prime.

Euclid's theorem on perfect numbers   *If p is a prime of the form $2^n - 1$, then the number $2^{n-1}p$ is perfect.*

Proof   By the preceding remarks, the proper divisors of $2^{n-1}p$ are

$$1, 2, 2^2, \ldots, 2^{n-1}, p, 2p, 2^2 p, \ldots, 2^{n-2}p.$$

To find the sum of these we need to know that $1 + 2 + 2^2 + \cdots + 2^{n-1} = 2^n - 1$. This can be done by the formula for the sum of a geometric series or, more naively, by adding 1 to the left-hand side and "folding it up" to 2^n as follows:

$$1 + 1 + 2 + 2^2 + 2^3 + \cdots + 2^{n-1}$$
$$= 2 + 2 + 2^2 + 2^3 + \cdots + 2^{n-1}$$

$$= 2^2 + 2^2 + 2^3 + \cdots + 2^{n-1}$$
$$= 2^3 + 2^3 + \cdots + 2^{n-1}$$
$$\vdots$$
$$= 2^{n-1} + 2^{n-1}$$
$$= 2^n$$

But now $2^n - 1 = p$, so when we add this to the other proper divisors the sum continues to fold up:

$$p + p + 2p + 2^2 p + 2^3 p + \cdots + 2^{n-2} p$$
$$= 2p + 2p + 2^2 p + 2^3 p + \cdots + 2^{n-2} p$$
$$= 2^2 p + 2^2 p + 2^3 p + \cdots + 2^{n-2} p$$
$$= 2^3 p + 2^3 p + \cdots + 2^{n-2} p$$
$$\vdots$$
$$= 2^{n-2} p + 2^{n-2} p$$
$$= 2^{n-1} p,$$

which is the number we started with. □

Exercises

Euclid's theorem shifts the focus of attention from perfect numbers to primes of the form $2^n - 1$. These are called *Mersenne primes* (as mentioned in connection with Exercise 1.2.4) because Mersenne recognized that they are prime only for prime n, and boldly conjectured that $n = 2, 3, 5, 7, 13, 17, 19, 31, 67, 127, 257$ give primes and $n = 89, 107$ do not. His conjectures were far from correct but were nevertheless important because they inspired Fermat to devise methods for finding factors of numbers of the form $2^n - 1$. Fermat's ideas turned out to be useful far outside this special problem, as we shall see in Chapter 6.

Although Euclid did not explicitly state unique prime factorization, there is evidence that the Greeks were aware of it and even of its implications for the description of divisors. Plato pointed out, in his *Laws* around 360 BC, that 5040 is a convenient number because it is divisible by all numbers from 1 to 10. He also mentioned that it has 59 divisors altogether.

The number of divisors is correct (if 5040 itself is omitted) and would be very hard to check except by using the fact that $5040 = 2^4 \times 3^2 \times 5 \times 7$.

1.7.1. Use the prime factorization of 5040 to show that it has $5 \times 3 \times 2 \times 2 = 60$ natural number divisors (including itself).

1.7.2. Show that $n = p_1^{e_1} p_2^{e_2} \cdots p_k^{e_k}$ has $(e_1 + 1)(e_2 + 1) \cdots (e_k + 1)$ natural number divisors.

Before leaving the subject of perfect numbers, it is worth mentioning that Leonhard Euler proved a converse of Euclid's theorem: *every even perfect number is of the form $2^{n-1}p$, where $p = 2^n - 1$ is prime.* An elegant proof of Euler's theorem, due to Leonard Eugene Dickson (1874–1934), goes as follows.

1.7.3. For any natural number $N = 2^{n-1}q$, where q is odd, let Σ be the sum of all natural number divisors of q. Show that the sum of all proper divisors of N is $(2^n - 1)\Sigma - N$.

1.7.4. Deduce from Exercise 1.7.3 that, if N is perfect, then $2N = 2^n q = (2^n - 1)\Sigma$ and hence $\Sigma = q + q/(2^n - 1)$.

1.7.5. Deduce from Exercise 1.7.4 that $2^n - 1$ divides q, that q and $q/(2^n-1)$ are the only divisors of q, and hence that q is a prime with $q = 2^n - 1$.

It remains an open problem whether there are any odd perfect numbers.

1.8   Induction

We began this book by claiming that arithmetic rests on the counting process and that proofs in arithmetic draw their strength from the logical essence of counting, *mathematical induction.* We gave one version of induction, called *descent,* and a few examples, and then said no more about it. So you may wonder whether induction is actually as important as we claimed. It is. Induction has been quietly intervening at crucial moments ever since we first mentioned it.

Look again over the previous sections, and you will see that descent was used to prove the following fundamental results:

• The division "algorithm" (or property) (Section 1.2).

• Existence of a prime divisor (Section 1.3).

- Termination of the Euclidean algorithm (Section 1.5).

- Unique prime factorization (Section 1.6).

It is also needed for Exercise 1.1.4 on Egyptian fractions, and Exercise 1.5.4 on $\gcd(F_{n+1}, F_n)$.

In addition to descent, which says that any descending sequence of natural numbers has a least member, we have used a form of induction that could be called *ascent*: if a sequence of natural numbers includes 1, and includes $i + 1$ when it includes i, then the sequence includes all natural numbers. This principle is immediate from the definition of the natural numbers by counting.

Ascent is normally used to prove a statement about n, $S(n)$ say, by proving that the sequence of numbers n for which $S(n)$ holds includes all natural numbers. One has to prove

1. $S(n)$ is true for $n = 1$
   (the so-called *base step*) and

2. $S(n)$ is true for $n = i + 1$ when it is true for $n = i$
   (the so-called *induction step*).

Then it follows by ascent that $S(n)$ is true for all natural numbers n. This form of induction was used in two crucial results.

- Correctness of the Euclidean algorithm (Section 1.5). To do this, we proved the statement S_n: $\gcd(a_n, b_n) = \gcd(a, b)$. It is true for $n = 1$, because $(a_1, b_1) = (a, b)$; and it is true for $n = i + 1$ when it is true for $n = i$, because $\gcd(a_i, b_i) = \gcd(a_{i+1}, b_{i+1})$.

- $\gcd(a, b) = ax + by$ for some integers x and y (Section 1.5). We actually proved the statement that a_n and b_n are of this form: proving it true for $n = 1$, because $a_1 = a$ and $b_1 = b$; then proving it true for $n = i + 1$ when it is true for $n = i$, because differences of numbers of the form $ax + by$ are still of this form.

Exercises

The ascent form of induction is often used to prove equations involving a sum of n terms, such as $S(n) : 1 + 2 + \cdots + n = \frac{n(n+1)}{2}$.

1.8.1. For this particular equation $S(n)$, check the base step $S(1)$. Then add $(i+1)$ to both sides of $S(i)$ to prove the induction step $S(i) \Rightarrow S(i+1)$.

1.8.2. Similarly use induction to prove that

$$1^2 + 2^2 + \cdots + n^2 = \frac{n(n+1)(2n+1)}{6}$$

and

$$1^3 + 2^3 + \cdots + n^3 = \left(\frac{n(n+1)}{2}\right)^2 = (1 + 2 + \cdots + n)^2.$$

On the other hand, a frequent complaint about such proofs is that one has to guess the right-hand side correctly before it is possible to get started. One would prefer a method that *discovers* the right-hand side, as well as proves it. For example, one can discover the form of $1 + 2 + \cdots + n$ by writing it a second time, in reverse:

$$1 + 2 + \cdots + (n-1) + n$$

$$n + (n-1) + \cdots + 2 + 1.$$

It is then clear, by adding the two rows, that there is a sum of $n+1$ in each of the n columns, hence

$$2(1 + 2 + \cdots + n) = n(n+1),$$

and therefore

$$1 + 2 + \cdots + n = \frac{n(n+1)}{2}.$$

The latter kind of proof is often called *noninductive*, but what has really happened is that induction has been redeployed to prove that each column has sum $n+1$. This is so easy that the base step and induction step need not be spelled out.

1.8.3. Use induction directly to prove the formula for the geometric series:

$$1 + r + r^2 + r^3 + \cdots + r^n = \frac{1 - r^{n+1}}{1 - r},$$

and describe a proof that leads to the discovery of this formula.

According to Hasse (1928), Zermelo found an interesting inductive proof of unique prime factorization along the following lines. Assuming there is a natural number with two different prime factorizations, there is a *least* such number n, by descent. It follows that n has two prime factorizations with no common prime factor, otherwise we could cancel to get a smaller number with two prime factorizations. Next ...

1.8.4. Suppose that p is a prime in the first factorization of n and q is a prime in the second. Show that pq does not divide n, otherwise there would be a smaller number with two different prime factorizations.

We now let $p < q$ be any two primes dividing the hypothetical least n with two prime factorizations. The following exercises derive a contradiction by showing that pq *does* divide n.

1.8.5. $\frac{n}{p} - \frac{n}{q} = \frac{n}{pq}(q-p)$ is a natural number. (Why?) Deduce that $\frac{n}{q}(q-p)$ is a natural number $< n$, hence with unique prime factorization, and that p divides $\frac{n}{q}(q - p)$.

1.8.6. But p does not divide $q - p$. (Why not?) Deduce from the unique prime factorization of $\frac{n}{q}(q - p)$ that p divides $\frac{n}{q}$, and hence that pq divides n, as required.

1.9 *   Foundations

The aim of mathematics is to prove things, which is hard, so mathematicians continually search for clearer and more powerful methods of proof. From time to time, this leads to criticism of existing methods as being unclear, or too complicated, or too narrow. Attempts are then made to find methods to replace them, which may lead to some parts of mathematics being rebuilt on different foundations. Historically, most of the rebuilding has been in the foundations of geometry and calculus, which we'll look at later, but in the 19th and 20th centuries it went as far as the foundations of arithmetic. The new foundations of arithmetic did not make the old ones obsolete, because in practice one gets along fine using induction and ring properties of $\mathbb{Z}$. But they were a revelation all the same, because they showed why induction is crucial to arithmetic: the ring properties can be derived from it. Thus arithmetic is *entirely* about the implications of the counting process!

This surprising discovery, which had been missed by all mathematicians from Euclid to Gauss, is due mainly to Hermann Grassmann (1861) and Richard Dedekind (1888).

Grassmann made the breakthrough by noticing that induction can be used not only as a method of proof, but as a method of *defi-*

nition. To define a function f on the natural numbers by induction, one writes down a value of $f(1)$, and a definition of $f(i+1)$ in terms of $f(i)$. It then follows by induction that $f(n)$ is defined for any natural number n. One function can be regarded as given along with the natural numbers themselves—the *successor* function $f(n) = n + 1$. All other standard functions, as Grassmann and Dedekind found, can be defined by induction.

In particular, Grassmann found that $+$ and $\times$ can be defined by induction, as follows. The defining equations for $+$ are

$$m + 1 = m + 1 \qquad \text{for all } m \tag{1}$$

$$m + (i + 1) = (m + i) + 1 \qquad \text{for all } m, i. \tag{i+1}$$

These equations are not as empty as they look! Equation (1) defines $m + n$ for $n = 1$ (and all natural numbers m) as the successor of m. Equation $(i+1)$ defines $m + (i+1)$ as the successor of $m + i$ (again for all natural numbers m). Thus the set of n for which $m + n$ is defined includes 1, and it includes $i + 1$ when it includes i, hence it includes all natural numbers, by induction.

Once $+$ is defined, $\times$ is defined inductively by the equations

$$m \times 1 = m \qquad \text{for all } m \tag{1}$$

$$m \times (i + 1) = m \times i + m \qquad \text{for all } m, i, \tag{i+1}$$

because the value of $m \times (i + 1)$ is defined in terms $m \times i$ and the previously defined function $+$.

The advantage of defining $+$ and $\times$ this way is that their properties can also be *proved* by induction. With suitable definitions of 0 and the negative integers, similar to those in Section 1.3, Grassmann found inductive proofs of all the ring properties of $\mathbb{Z}$ (see the exercises). Thus induction is a complete foundation for arithmetic.

Richard Dedekind (1888) asked himself the question: what are the properties of the successor function that allow it to serve as a basis for the rest of arithmetic? He found the answer to this question in terms of *sets*—a radical idea at the time, but one that has since been accepted as the most reasonable way to provide a foundation for all of mathematics.

The essential properties of the successor function are very simple.

1. It is defined on an infinite set (namely, the set of natural numbers).

2. It is one-to-one (that is, unequal numbers have unequal successors).

3. It is not onto the whole set (in particular, 1 is not a successor).

Dedekind realized that *any* function f with these properties gives rise to a set that "behaves like" the natural numbers. If a is an element that is not a value of f, then $a, f(a), f(f(a)), \ldots$ behave like $1, 2, 3, \ldots$. Thus the *abstract structure* of the natural numbers is completely described by an infinite set and a function that is one-to-one but not onto.

However, it is a deep philosophical question whether infinite sets can actually be proved to exist. Dedekind gave an answer that is very interesting, although it lies outside mathematics. He said that such a set is the set of possible ideas, because for every idea I there is another idea $f(I) =$ the idea of I, which is distinct from I. Indeed the "idea of" function f behaves like a successor function on the set of ideas.

Mathematicians have not accepted Dedekind's set of ideas as a genuine set, and the existence of infinite sets is taken as an axiom, so we will not attempt to prove it. However, the statement of this *axiom of infinity* (as it is called) is remarkably similar to Dedekind's description of the set of ideas. For each set X we define a "successor" of X by taking X as a member of a new set $\{X\}$ (rather like forming the "idea of X"). The actual successor of X is taken to be $X \cup \{X\}$, the set whose members are the members of X *and* X itself, for technical reasons. Then the axiom of infinity says that there is a set Ω rather like the set of ideas: Ω is not empty (in fact, take the empty set to be one of its members), and along with each X in Ω, the successor of X is also in Ω.

Thus when we pursue the natural numbers into the depths of set theory, what we find is nothing but the empty set and its successors. But this is all we need! John von Neumann (1923) suggested that this is the best way to define the natural numbers, or rather, the natural numbers together with zero, because the empty set is surely the best possible set to represent zero. Here is what the first few numbers look like, according to his definition.

$$0 = \{\} \qquad \text{(the empty set)}$$
$$1 = \{0\} \qquad \text{(the set whose only member is 0)}$$
$$2 = \{0, 1\}$$
$$3 = \{0, 1, 2\}$$
$$\vdots$$

In other words, 0 is the empty set, and each natural number is the set of its predecessors. You must admit that this definition can hardly be beaten for economy, because everything is built out of "nothing"—the empty set. It is also quite natural and elegant, because the ordering of natural numbers is captured by membership, the basic concept of set theory: $m < n \Leftrightarrow m$ is a member of n. Last but not least, von Neumann's definition is a very snappy answer if anyone ever forces you to give a definition of the natural numbers!

Exercises

You may have noticed that the second equation in the definition of $+$, namely,

$$m + (i + 1) = (m + i) + 1,$$

is a special case of the associative law for $+$. In fact, this was precisely Grassmann's starting point in his inductive proof of the ring properties of $\mathbb{Z}$. The associative law for $+$ (in $\mathbb{N}$) may be formulated as a statement about n by letting $S_1(n)$ be the statement:

$$l + (m + n) = (l + m) + n \qquad \text{for all natural numbers } l \text{ and } m.$$

1.9.1. Show that $S_1(1)$ is true by definition of $+$.

1.9.2. Prove $S_1(i) \Rightarrow S_1(i + 1)$ with the help of $S_1(1)$.

Grassmann's next goal was to use associativity of $+$ to prove commutativity of $+$ in $\mathbb{N}$, again inductively. However, it is not even clear that $1 + n = n + 1$, so the latter statement, call it $S_2(n)$, must be proved first. $S_2(1)$ is $1 + 1 = 1 + 1$, so $S_2(1)$ is true!

1.9.3. Prove $S_2(i) \Rightarrow S_2(i + 1)$ using associativity of $+$.

Finally, we can let $S_3(n)$ be the full commutativity statement for $\mathbb{N}$:

$$m + n = n + m \qquad \text{for all natural numbers } m.$$

$S_3(1)$ is $1 + n = n + 1$, which has just been proved, so it remains to do the following.

1.9.4. Prove $S_3(i) \Rightarrow S_3(i + 1)$ using associativity of $+$ and $S_3(1)$.

Now let us switch to Dedekind's work. When we said that $a, f(a), f(f(a)), \ldots$ "behave like" $1, 2, 3 \ldots$, you may have wanted to ask: what is the exact meaning of the three dots? This is a fair question, because f could be defined beyond where we intend the sequence $a, f(a), f(f(a)), \ldots$ to go. Here is an example.

1.9.5. Let S be the union of $\mathbb{N}$ with the set $\mathbb{Z} + 1/2 = \{m + 1/2 : m \in \mathbb{Z}\}$, and let $f(x) = x + 1$. Now show

(a) f is defined for all members of S, is one-to-one, but not onto S.

(b) S does *not* behave like $\mathbb{N}$, because infinite descent is possible in S.

In this example, $a = 1$, and the intended meaning of $\{a, f(a), f(f(a)), \ldots\}$ is the set $\mathbb{N}$. But how do we capture the meaning of "$\ldots$" in other cases without using expressions like "obtainable in a finite number of steps" which assume what we are trying to define? Dedekind also had an answer to this question. He said that $\{a, f(a), f(f(a)), \ldots\}$ consists of the elements that belong to *all* sets that include a, and that include $f(x)$ when they include x.

1.9.6. Ponder Dedekind's definition, and show that it also enables us to define $\mathbb{N}$ from the set Ω asserted to exist by the axiom of infinity.

1.10   Discussion

The Euclidean Algorithm

The Euclidean algorithm is a splendid example of the universality of mathematics. It seems to have been discovered in three different cultures and for several different mathematical purposes. In ancient Greece it was crucial in Euclid's theory of divisibility and primes, as we have seen, and it was also important in the study of irrational numbers (see Section 8.6*). Euclid's proof of the prime divisor property was actually more complicated than the one given in Section 1.6, because he apparently did not know that $\gcd(a, b) = ma + nb$ for integers m and n.

This linear representation of the gcd was discovered in India and China, perhaps first by Âryabhaṭa and Bhâskara I around 500 AD. The Indian mathematicians were interested in integer solutions of equations $ax + by = c$, and this depends on finding $\gcd(a, b)$ in the form $ma + nb$, as we saw in Section 1.5. Such problems also arose in Chinese mathematics, particularly in the so-called "Chinese remainder" problems we shall study in Section 6.6.

The algorithm became familiar in Europe by the 16th century, but for another 200 years it was considered just a useful tool rather than a revealing property of numbers. Gauss avoided use or mention of the Euclidean algorithm in his *Disquisitiones Arithmeticae*. He avoided using it for the fundamental theorem of arithmetic by giving a direct proof, by descent, of the prime divisor property (the one covered in Exercises 1.6.4 and 1.6.5). He did not even mention it when discussing the gcd and lcm, giving instead the rules for computing them from prime factorizations, and saying only that

> we know from elementary considerations how to solve these problems when the resolution of the numbers A, B, C, etc. into factors are not given (*Disquisitiones*, article 18).

And he hid its role in the solution of $ax + by = 1$ (article 28) by referring only to the so-called "continued fraction" method, which is equivalent.

Dirichlet simplified, and in some ways extended, the *Disquisitiones* in his *Vorlesungen über Zahlentheorie* (lectures on number theory) of 1867. One of his reforms was reinstatement of the Euclidean algorithm. He used it to derive the fundamental theorem and related results much as we have in this chapter, and went so far as to say:

> It is now clear that the whole structure rests on a *single* foundation, namely the algorithm for finding the greatest common divisor of two numbers. All the subsequent theorems, even when they depend on the later concepts of relative and abso-

lute prime numbers, are still only simple consequences of the result of this initial investigation ... (Dirichlet (1867), §16).

One of the reasons Dirichlet was enthusiastic about the Euclidean algorithm was that it could be used in other situations, a fact that also converted Gauss in the end. In 1831, Gauss found it useful to introduce what are now called *Gaussian integers* —numbers of the form $a+b\sqrt{-1}$, which we shall study in Chapter 7—and found that the key to their arithmetic was the applicability of the Euclidean algorithm. Perhaps it was with this generalization in mind that Dirichlet based his number theory on the Euclidean algorithm from the beginning, because the passage quoted above continues:

> ... so one is entitled to make the following claim: any analogous theory, for which there is a similar algorithm for the greatest common divisor, must also have consequences analogous to those in our theory.

Induction

The ascent form of induction is now considered indispensible in all fields of mathematics that use natural numbers, so one would expect to find it in the earliest mathematical works. Surprisingly, it does not seem to be there. The first clear statement of the "base step, induction step" format first appeared in 1654. How did mathematicians manage for so long without this essential tool?

The answer, I believe, is that until recently mathematicians preferred *descent* to the "base step, induction step" form of induction we called *ascent* in Section 1.8. Descent is not only simpler than ascent because no "base step" is involved; it also seems to occur more naturally at the lower levels of mathematics.

Examples of descent date from ancient times, at least as far back as Euclid's *Elements*, around 300 BC, and conceivably in proofs that $\sqrt{2}$ is irrational. Euclid uses descent in Proposition 31 of Book VII of the *Elements*, to prove that any composite number A has a prime divisor or, as he puts it, that A is "measured" by some prime. He argues that A has some divisor B, because A is composite. If B is prime we are done; if not, B has a divisor C, and so on. He then claims

> Thus, if the investigation be continued in this way, some prime number will be found which will measure the number before it, which will also measure A.

And his punchline is an appeal to descent:

> For, if it is not found, an infinite series of numbers will measure the number A, each of which is less than the other: which is impossible in numbers.

Euclid also assumes termination of the Euclidean algorithm without comment throughout the *Elements*. It *is* obvious, of course, but hardly more obvious than the existence of a prime divisor. Evidently Euclid was only fleetingly aware of the importance of induction; nevertheless it is to his credit that he noticed it at least once.

Most mathematicians failed to notice descent until around 1640, when Fermat began to announce spectacular new results in number theory and claim they were due to a "method of infinite descent." His most famous proof, and in fact the only one he disclosed, shows that there are no natural numbers a, b, and c such that $a^4 + b^4 = c^2$. He assumed, on the contrary, that there is a solution $a = x_1$, $b = y_1$, $c = z_1$, and showed how to descend to a *smaller* solution $a = x_2$, $b = y_2$, $c = z_2$. By descending indefinitely in this way, one obtains a contradiction that proves the desired result "by infinite descent." The details may be seen in Section 4.7*.

This proof made mathematicians conscious of descent for the first time and hinted at its power. At the same time, unfortunately, the simple logical principle of descent was buried under the technical problem of finding the descent step. Mathematicians continued to use descent until the late 19th century without realizing that an important principle was involved. The Gauss (1801) proof of the prime divisor property (Exercises 1.6.4 and 1.6.5) is a simpler example.

Ascent was likewise used for a long time without the importance of the induction step being noticed. Mathematicians naturally tried to make proofs as simple as possible, so ascent proofs were organized to make the induction step trivial, and hence not worth mentioning. A brilliant example is Euclid's summation of the geometric series (*Elements*, Book IX, Proposition 35). It was also easier to *discover* results in these circumstances. And as long as it was possible to play down the induction step, it was possible to overlook the underlying principle of induction.

The induction step ultimately came to light not in number theory but in combinatorics, where complicated inductions perhaps arise more naturally. The first really precise formulation of induction is by Blaise Pascal (1654), who clearly used the "base step, induction step" format to prove the basic properties of Pascal's triangle.

Understanding did not advance much between 1654 and 1861. Ascending and descending forms of induction were both occasionally used, but without recognition of their importance, or even their equivalence. Certainly, one would not think the time was ripe for a high school teacher to write a textbook using mathematical induction as the *sole basis* of arithmetic! Enter Hermann Grassmann. His *Lehrbuch der Arithmetik für höhere Lehranstalten* (textbook of arithmetic for higher instruction) contains the fundamental idea that everyone else had missed: *the whole of arithmetic follows from the process of succession*. As we explained in Section 1.9*, he did this by using induction to define $+$ and $\times$ from the successor function, and hence prove the ring properties of $\mathbb{Z}$.

But, sadly, Grassmann was a generation ahead of his time. His work fell into obscurity so fast that even like-minded mathematicians of the 1880s and 1890s were unaware of it. Dedekind (1888) rediscovered the inductive definitions of $+$ and $\times$ in terms of the successor function and decided to dig deeper, to explain the nature of *succession* itself. As we asked before Exercise 1.9.5: in the expression $1, 2, 3, \ldots$, what does $\ldots$ mean? It is not enough to say "the remaining values of the successor function $f(n) = n + 1$," because $f(n)$ is also defined on the numbers $n = m + 1/2$ for integers m, and we do not intend these values to be included among the successors of 1. The crux of the problem of defining succession is to exclude such "alien intruders," as Dedekind called them.

As indicated in Section 1.9*, Dedekind's solution makes crucial use of set theory. His discoveries were in fact very influential in the development of logic and set theory in the 20th century. Also, his (and Grassmann's) method of definition by induction led to the theory of recursive functions, and ultimately to computer programming and computer science. This is a surprising twist to a basically philosophical investigation, but mathematics often seems to find its way into the real world, without being asked.

2 Geometry

2.1 Geometric Intuition

Geometry is in many ways opposite or complementary to arithmetic. Arithmetic is discrete, static, computational, and logical; geometry is continuous, fluid, dynamic, and visual. The fundamental geometric quantities (length, area, and volume) are familiar to everyone but hard to define. And some "obvious" geometric facts are not even provable; they can be taken as axioms, but so can their opposites. In geometry, intuition runs ahead of logic. Our imagination leads us to conclusions via steps that "look right" but may not have a purely logical basis. A good example is the Pythagorean theorem, that the square on the hypotenuse of a right-angled triangle equals (in area) the sum of the squares on the other two sides. This theorem has been known since ancient times; was probably first noticed by someone playing with squares and triangles, perhaps as in Figure 2.1.

The picture on the left shows a big square, minus four copies of the triangle, equal in area to the squares on the two sides. The second picture shows that the big square minus four copies of the triangle also equals the square on the hypotenuse. Q.E.D.

This is a wonderful discovery (and it gets better, as we shall see later), but what is it really about? In the physical world, exact

37

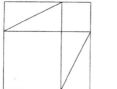

FIGURE 2.1   A proof of the Pythagorean theorem.

triangles and squares do not exist, so the theorem has to be about some kind of ideal or abstract objects. And yet, we are surely using our experience with actual triangles and squares to draw conclusions about the abstract ones.

Thus the gift of geometric intuition is both a blessing and a curse. It gives us amazingly direct access to mathematical results; yet we cannot be satisfied with the results seen by our intuition until they have been validated by logic. The validation can be very hard work, and it would be disappointing if its *only* outcome was confirmation of results we already believe. A method of validating intuition is worthwhile only if it takes us further than intuition alone.

The most conservative solution to the problem of validating intuition is the so-called *synthetic geometry*. In this system, all theorems are derived by pure logic from a (rather long) list of visually plausible axioms about points, lines, circles, planes, and so on. Thus we can be sure that all theorems proved in synthetic geometry will be intuitively acceptable. This was the approach initiated in Book I of Euclid's *Elements* and perfected in David Hilbert's *Foundations of Geometry* (1899). Its advantages are that it is self-contained (no concepts from outside geometry) and close to intuition (the steps in a proof may imitate the way we "see" a theorem). However, it fails to explain the mysterious similarity between geometry and arithmetic; the fact that geometric quantities, like numbers, can be added, subtracted, and (in the case of lengths) even multiplied. It looks like geometry and arithmetic share a common ground, and mathematics should explain why.

The search for a common ground of arithmetic and geometry led to the so-called *analytic geometry*, initiated in René Descartes' *Geometry* (1637) and also perfected by Hilbert. It is more efficient as a way of making geometry rigorous, and history has shown it to be more fruitful than synthetic geometry in its consequences. It enriches both arithmetic and geometry with new concepts, and in fact with the whole new mathematical world of algebra and calculus. As we shall see in later chapters, the new world is not separate from the old, but it increases our understanding of it. Algebra not only throws new light on geometry, it also enables us to solve problems about the natural numbers that were previously beyond reach.

Analytic geometry will be developed in the next chapter. In the meantime, we will use intuition freely to gain a bird's eye view of the landmark results and concepts in geometry, to see how far arithmetic concepts apply to geometric quantities, and to see why the number concept needs to be extended to build a common foundation for arithmetic and geometry.

Exercises

The Pythagorean theorem has been discovered many times, in different cultures, and proved in many different ways. The very immediate proof indicated in Figure 2.1 was given by Bhâskara II in 12th century India. Another way the theorem may have been discovered was suggested by Magnus (1974), p. 159. It comes from thinking about the tiled floor shown in Figure 2.2.

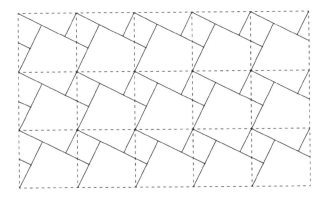

FIGURE 2.2   Pythagorean theorem in a tiled floor.

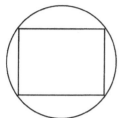

FIGURE 2.3   Rectangle in a circle.

2.1.1. Explain how Figure 2.2 is related to the Pythagorean theorem. (The dotted squares are not the tiles; they are a hint.)

The converse Pythagorean theorem is also important: *if a, b, and c are lengths such that $a^2 + b^2 = c^2$ then the triangle with sides a, b, and c is right-angled* (with the right angle formed by the sides *a* and *b*).

2.1.2. Deduce the converse Pythagorean theorem from the Pythagorean theorem itself.

2.1.3. Deduce the Pythagorean theorem from its converse.

Another very old theorem is that *any right-angled triangle fits in a semicircle, with the hypotenuse as diameter.* According to legend, synthetic geometry began with a proof of this theorem by Thales in the 6th century B.C.

2.1.4. Why might Figure 2.3 lead you to believe that any right-angled triangle fits in a semicircle?

2.2   Constructions

The aim of Euclid's geometry is to study the properties of the simplest curves, the straight line, and the circle. These are drawn by the simplest drawing instruments, the ruler and the compass; hence much of the *Elements* consists of so-called *ruler and compass constructions*. In fact, two of Euclid's axioms state that the following constructions are possible.

- To draw a straight line from any point to any other point.
- To draw a circle with any center and radius.

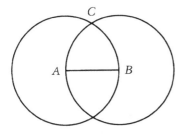

FIGURE 2.4 Constructing an equilateral triangle.

His axioms do not state the existence of anything else, so all other figures are shown to exist by actually constructing them. Certain points, lines, and circles being given, new ones are constructed using ruler and compass. This creates new points, from which new lines and circles are constructed, and so on, until the required figure is obtained.

Euclid's first proposition is that it is possible to construct an equilateral triangle with a given side AB, and his first figure shows how it is done (Figure 2.4). Namely, draw the circle with center A and radius AB, then the circle with center B and radius BA, and connect A and B by straight lines to one intersection, C, of these circles.

Several other important constructions come from this.

1. **Bisecting an angle.**
   Drawing a circle with center at the apex O of the angle marks off equal sides OA and OB (Figure 2.5). Then if we construct an equilateral triangle ABC, the line OC will bisect the angle AOB.

2. **Bisecting a line segment.**
   Given the line segment AB, construct the equilateral triangle ABC. Then the bisector of the angle ACB also bisects AB.

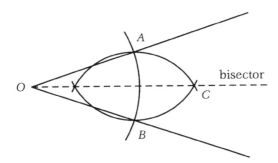

FIGURE 2.5 Bisecting an angle.

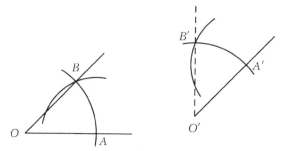

FIGURE 2.6
Replicating an angle.

3. Constructing the perpendicular to a line through a point O not on it.

Draw a circle with center O, large enough to cut the line at points A and B. Then the bisector of angle AOB will be the required perpendicular.

As one notices from these constructions, the compass gives an easy way to replicate a given line segment. It is also possible to replicate a given angle. For example, one can draw a circle with unit radius centered on the apex O of the angle, then use the line AB between its intersections as a second radius to find points A', B' so that angle $B'O'A' = $ angle BOA, with O' a given point on the given line (Figure 2.6). Euclid uses angle replication to construct a parallel to a given line through a given point. He chooses a point O at random on the line, joins it to the given point O', then replicates the angle $O'OX$ as angle $OO'X'$ (Figure 2.7).

The construction of parallels is needed to divide a line segment AB into n equal parts, for any natural number n. (The special case

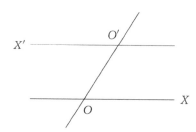

FIGURE 2.7   Constructing a parallel.

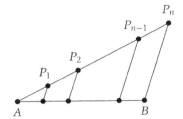

FIGURE 2.8 Cutting a line segment into equal parts.

$n = 2$, or bisection, is not typical, because it does not require parallels.) Along any line through A (other than AB) mark n equally spaced points $P_1, P_2, \ldots, P_n$ by repeatedly replicating an arbitrary line segment AP_1. Then join P_n to B, and draw parallels to $P_n B$ through $P_1, P_2, \ldots, P_{n-1}$ (Figure 2.8). These parallels cut AB into n equal parts.

Exercises

Most of the following problems are variations on the theme of finding the perpendicular bisector of a line segment.

2.2.1. Describe the construction of the perpendicular to a line through a given point on the line.

2.2.2. Given a circle, but *not* its center, give a construction to find the center.

2.2.3. The perpendicular bisectors of the sides of a triangle meet at a single point. What property of the perpendicular bisector makes this obvious?

2.2.4. Use Exercise 2.2.3 to find a circle passing through the vertices of any triangle.

2.2.5. Describe the construction of a square and a regular hexagon.

The Greeks also found a construction of the regular pentagon, but no essentially new constructions were found until Gauss in 1796 found how to construct the regular 17-gon. This led to an algebraic theory of constructibility that explained why no constructions had been found for the regular 7-gon, 11-gon, and others. The astonishing result of Gauss's theory (completed by Pierre Wantzel in 1837) is that the regular n-gon is constructible if and only if n is the product of a power of 2 by distinct Fermat primes. (Recall that these were defined in Exercise 1.2.7.)

2.2.6. If $\gcd(m, n) = 1$ and the regular m-gon and n-gon are constructible, show that the regular mn-gon is also constructible.

2.3   Parallels and Angles

The crucial assumption in Euclid's geometry—the one that makes the geometry "Euclidean"—is the *parallel axiom*. It can be stated in many different ways, the most concise of which is probably the one given by Playfair in 1795.

Parallel axiom. *If $\mathcal{L}$ is a line and P is a point not on $\mathcal{L}$, then there is exactly one line through P that does not meet $\mathcal{L}$.*

The single line through P that does not meet $\mathcal{L}$ is called the *parallel* to $\mathcal{L}$ through P. Euclid's statement is more complicated, and it involves the concept of angle, which is not mentioned in the Playfair version. This would ordinarily be regarded as inelegant mathematics, but in this case it is more informative, and it points us toward some important consequences of the parallel axiom.

That, if a straight line falling on two straight lines make the interior angles on the same side less than two right angles, the two straight lines, if produced indefinitely, meet on that side on which are the angles less than the two right angles. [From the edition of Euclid's *Elements* by Heath (1925), p. 202]

Figure 2.9 shows the situation described by Euclid, which is what happens with *non*parallel lines. If the angles α and β have a sum less than two right angles, then $\mathcal{L}$ meets $\mathcal{M}$ on the side where α and β are. To see why Euclid's statement is equivalent to Playfair's, one only needs to know that angles α and β sum to two right angles if they can be moved so that together they form a straight line (Figure 2.10). The same figure shows that the *vertically opposite angles*, both marked α, are equal, because each of them plus β equals two right angles.

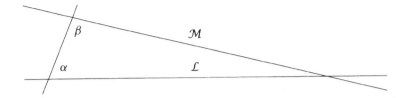

FIGURE 2.9   Nonparallel lines.

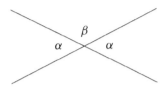

FIGURE 2.10   Vertically opposite angles.

Euclid's statement of the parallel axiom tells us that a line $\mathcal{M}$ can fail to meet line $\mathcal{L}$ only when the two interior angles α and β sum to two right angles. This follows from the facts about angles just mentioned. If $\alpha + \beta$ is greater than two right angles, then the interior angles on the *other* side of the transverse line $\mathcal{N}$ will sum to less than two right angles, and hence $\mathcal{M}$ will meet $\mathcal{L}$ on the other side. If $\alpha + \beta$ equals two right angles, the interior angles on the other side of the transverse line $\mathcal{N}$ are also α and β, hence also of sum equal to two right angles. In this case it follows by symmetry that $\mathcal{L}$ and $\mathcal{M}$ meet on both sides or neither side. The possibility of two meetings is ruled out by another axiom, that there is only one straight line through any two points. Hence there is exactly one line $\mathcal{M}$ through P that does not meet $\mathcal{L}$: the line for which $\alpha + \beta$ equals two right angles (Figure 2.11). The most important property of angles that follows from the parallel axiom is that *the angle sum of a triangle is two right angles*. The proof is based on Figure 2.12, which shows an arbitrary triangle with a parallel to one side drawn through the opposite vertex. The angles of the triangle recur as shown and hence sum to a straight line.

It follows, by pasting triangles together, that the angle sum of any quadrilateral is four right angles. In particular, in a quadrilateral with equal angles, each angle is a right angle. This means that

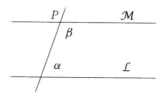

FIGURE 2.11   Parallel lines and related angles.

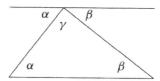

FIGURE 2.12   Angle sum of a triangle.

rectangles and squares of any size exist. Of course, this is what we always thought, but we can now see that it follows from a small number of more basic statements about straight lines, among them the crucial parallel axiom. As we shall see in Section 2.5, the existence of rectangles is the key to the intuitive concept of area and to finding a common ground for geometry and arithmetic.

Exercises

The crucial role of the parallel axiom can be seen from the number of important statements that are equivalent to it, and hence cannot be proved without it. Not only are the Playfair and Euclid versions equivalent to each other (Exercise 2.3.1 completes the proof of this), they are also equivalent to the statement about the angle sum of a triangle (Exercise 2.3.2).

2.3.1. Deduce Euclid's version of the parallel axiom from Playfair's.

2.3.2. Deduce Euclid's version of the parallel axiom from the statement that the angle sum of a triangle is two right angles.

2.3.3. Assuming that any polygon can be cut into triangles, show that the angle sum of any n-gon is $(n-2)\pi$, where π denotes two right angles.

2.3.4. Deduce from Exercise 2.3.3 that the only ways to tile the plane with copies of a single regular n-gon (that is, an n-gon with equal sides and equal angles) are by equilateral triangles, squares, and regular hexagons.

2.3.5. Show that the plane can be tiled with copies of any single triangle.

2.4 Angles and Circles

One of the first theorems in Euclid's *Elements* says that *the base angles of an isosceles ("equal sides") triangle are equal.* The most elegant proof of this theorem was found by another Greek mathematician, Pappus, around 300 A.D. It goes like this. Suppose ABC is a triangle with $AB = BC$ (Figure 2.13). Because $AB = BC$, this triangle can be turned over and placed so that BC replaces AB, and AB replaces BC. In other words, the triangle exactly fills the space it filled in its old position. In particular, the base angle BAC fills the space previously filled by angle BCA, so these two angles are equal.

Triangles that occupy the same space were called *congruent* by Euclid. He used the idea of moving one triangle to coincide with another to prove the two triangles congruent when they agree in certain angles and sides. The preceding argument uses "side-angle-side" agreement: if two triangles agree in two sides and the included angle, then one can be moved to coincide with the other. Congru-

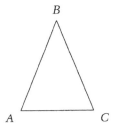

FIGURE 2.13 An isosceles triangle.

ence also occurs in the angle-side-angle and side-side-side cases. Later mathematicians felt that the idea of motion did not belong in synthetic geometry, and instead stated the congruence of triangles with side-angle-side, angle-side-angle, or side-side-side agreement as axioms. This was done in Hilbert's *Foundations of Geometry* (1899), for example. The idea of motion came back in Felix Klein's definition of geometry, which we shall discuss in Chapter 3.

Whichever approach is adopted, the theorem on the base angles of an isosceles triangle is the key to many other results. Perhaps the most important is the theorem relating angles in a circle: *an arc of a circle subtends twice the angle at the center as it does at the circumference.* Figure 2.14 shows the situation in question—the arc AB and the angles AOB and APB it subtends at the center and circumference, respectively—together with a construction line PQ, which gives away the plot.

Because the lines OA and OP are radii of the circle, they are equal. Therefore, triangle POA is isosceles, with equal base angles α as shown. The external angle QOA is therefore 2α because it, like the interior angles α, forms a sum of two right angles with the interior angle AOP. Similarly, the triangle POB has equal angles β as shown and an exterior angle 2β. Thus the angle $2(\alpha + \beta)$ at the center is twice the angle $\alpha + \beta$ at the circumference.

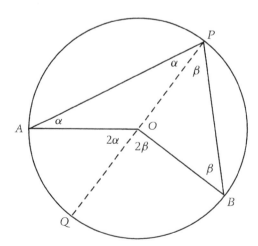

FIGURE 2.14 Angles subtended by an arc.

In the special case where the arc *AB* is half the circumference so that the angle at the center is straight, we find that the angle at the circumference is a right angle. This should remind you of the theorem of Thales mentioned in the exercises for Section 2.1.

Exercises

We originally stated the theorem of Thales by saying that any right-angled triangle fits in a semicircle. The special case of the theorem about angles in a circle says, rather, that any triangle in a semicircle is right-angled. The two theorems are actually converses of each other. However, they are both true, and the relationship between them can be traced back to converse theorems about isosceles triangles.

2.4.1. Prove that a triangle with two equal angles is isosceles.

2.4.2. What form of congruence axiom is involved in Exercise 2.4.1?

From Exercise 2.4.1, which is the converse theorem about isosceles triangles, we deduce the converse Thales' theorem. It is based on Figure 2.15.

2.4.3. If triangle *PAB* has a right angle at *P* and *PO* is drawn to make the equal angles marked α, show that this also results in equal angles marked β.

2.4.4. Deduce from Exercise 2.4.3 that each right-angled triangle fits in a semicircle.

The fact the angle at the circumference is half the angle at the center implies that the angle at the circumference is *constant*. This means that

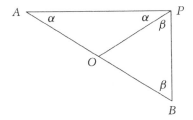

FIGURE 2.15 A right-angled triangle.

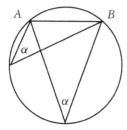

FIGURE 2.16   Apparent size of a chord of a circle.

the chord AB looks the same size, viewed from any point on the circle (Figure 2.16).

Now suppose that we vary the circle through A and B, and consider the effect on the apparent size of AB (a problem of practical importance if, say, you are trying to score a goal between goalposts A and B).

2.4.5. Show that the maximum apparent size of AB, viewed from a line CD, occurs at the point where CD is tangential to a circle through A and B.

2.5   Length and Area

Arithmetic and geometry come together in the idea of *measurement*, first for lengths, but more interestingly for areas. In fact, the very word "geometry" comes from the Greek for "land measurement." To measure lengths, we choose a fixed line segment as the unit of length and attempt to express other lengths as multiples of it. By joining copies of the unit end to end we can obtain any natural number multiple of the unit, and by dividing these into equal parts, we also obtain any rational multiple of the unit. For most practical purposes, this is sufficient, because rational multiples of the unit can be as small as we please. However, we know from Section 1.1 that $\sqrt{2}$ is not rational, and mathematicians would like to be able to speak of a length $\sqrt{2}$, even though the similar length 1.414 might be near enough for surveying or carpentry.

The fundamental problem in measurement is to find enough numbers to represent all possible lengths. This problem is more difficult than it looks, and we shall postpone it until the next chapter.

For the time being we shall just assume that every length is a number. The Greeks did not believe such an assumption was necessary, but this caused difficulties with their theory of area, as we shall see now and in the next section.

Just as length is measured by counting unit lengths, area is measured by counting unit *squares*, that is, squares whose sides are of unit length. For example, a rectangle of height 3 and width 5 can be cut into $3 \times 5 = 15$ unit squares, as Figure 2.17 clearly shows; hence it has area 15.

How convenient that we call it a 3×5 rectangle! Multiplication is the natural symbol to describe rectangles, because it gives the number of unit squares in them. And not only when the sides are integer multiples of the unit. A rectangle of height 3/2 and width 5/2 can similarly be cut into 15 squares of side 1/2, each of which has area 1/4 (because four of them make a unit square). Hence the area of the $3/2 \times 5/2$ rectangle is $3/2 \times 5/2 = 15/4$.

The same idea, cutting into little fractional squares, shows that the area of an $r \times s$ rectangle is rs for any rational multiples r and s of the unit. But what about, say, a square with side $\sqrt{2}$? Is its area $\sqrt{2} \times \sqrt{2} = 2$? Well, the area of an $r \times r$ square should be close to the area of a $\sqrt{2} \times \sqrt{2}$ square when r is a rational number close to $\sqrt{2}$. If so, 2 is the area of the $\sqrt{2} \times \sqrt{2}$ square, because 2 is the number approached by the values r^2 as r approaches $\sqrt{2}$.

For the Greeks, the area of the $\sqrt{2} \times \sqrt{2}$ square was not a problem, because they *defined* $\sqrt{2}$ to be the side of a square of area 2. The price they paid for this was having to develop a separate arithmetic of lengths and areas, since they did not regard $\sqrt{2}$ as a number. If one wants all lengths to be numbers, defining the area of a rectangle is the same as defining the product of irrational lengths, and it can only be done by comparing the rectangle with arbitrarily close rational rectangles. Once the area of a rectangle is known to be

FIGURE 2.17   A 3×5 rectangle.

FIGURE 2.18   Area of a parallelogram.

height × width, however, there is a simple way to find the area of other polygons: by *cutting and pasting*.

For example, the standard proof that the area of a triangle is $\frac{1}{2}$base×height is achieved by cutting and pasting. We first argue that the area of the triangle is half the area of the parallelogram obtained by pasting two copies of the triangle together, then that

$$\text{area of parallelogram} = \text{base} \times \text{height}$$

by cutting a triangle off one end of the parallelogram and pasting it on to the other to make a rectangle with the same base and height (Figure 2.18). After this, the area of any other polygon follows, because any polygon can be cut into triangles.

Exercises

It is not quite obvious that any polygon can be cut into triangles, so we should check that this is true, because it is the only way we know to define the area of a polygon.

2.5.1. A polygon Π is *convex* if the line segment connecting any two points of Π is contained in Π. Show that a convex polygon with n sides can be cut into n triangles.

Thus it now suffices to prove that any polygon can be cut into a finite number of convex polygons. This can be done in two easy steps.

2.5.2. Show that any finite set of lines divides the plane into convex polygons.

2.5.3. Deduce from Exercise 2.5.2 that any polygon is cut into convex pieces by the lines that extend its own edges.

2.6   The Pythagorean Theorem

Having seen the main ideas of Greek geometry, it is worth looking
again at the Pythagorean theorem, to see where it fits into the big
picture. Logically, it comes after the basic theory of area, and in fact
Euclid uses the fact that a triangle has half the area of a parallelogram
with the same base and height. His proof goes as follows (referring
to Figure 2.19).

> square *ABFG* on one side of the triangle
>
> $= 2 \times$ triangle *CFB*
>
> (same base and height),
>
> $= 2 \times$ triangle *ABD*
>
> (because the triangles are congruent by agreement
> of side-angle-side),
>
> $=$ rectangle *BMLD*
>
> (same base and height).

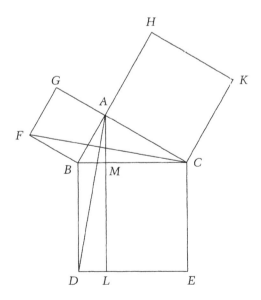

FIGURE 2.19   Areas related to the right-angled triangle.

Similarly, square $ACKH$ on the other side of the triangle equals rectangle $MCEL$, so the squares on the two sides sum to the square on the hypotenuse.   □

As mentioned in Section 2.1, the Pythagorean theorem was discovered in several different cultures, and in fact some of them discovered it long before the time of Pythagoras. However, Pythagoras and his followers (the *Pythagoreans*) deserve special mention because they also discovered that $\sqrt{2}$ is irrational. According to legend, this discovery caused great dismay because it conflicted with the Pythagorean philosophy that "all is number." The Pythagoreans initially believed that all things, including lengths, could be measured by natural numbers or their ratios. Yet they could not deny that the diagonal of the unit square was a length, and according to the Pythagorean theorem its square was equal to 2, hence the side and diagonal of the square were *not* natural number multiples of a common unit.

The first fruits of the conflict were bitter, to our taste, but they had a huge influence on the development of mathematics.

- Separation of arithmetic and geometry.

- Development of a separate arithmetic of lengths and areas.

- Preference for the latter "geometric" arithmetic, and the development of a corresponding "geometric algebra."

In geometric algebra, lengths are added by joining them end to end and multiplied by forming the rectangle with them as adjacent sides, the product being interpreted as its area. Areas are added by pasting. From the Pythagorean viewpoint, it is natural to relate lengths via areas. The basic example is the Pythagorean theorem itself, which says that the sides and hypotenuse of a right-angled triangle are simply related via their squares, even though they are *not* simply related as lengths.

The sweeter fruits of the conflict grew from the eventual reconciliation of arithmetic and geometry. This began only in the 17th century, when Fermat and Descartes introduced analytic geometry, and it was not completed until the late 19th century. It was difficult because:

- Arithmetic and geometry involve very different styles of thought, and it was not clearly possible or desirable to do geometry "arithmetically."

- Defects in Euclid's geometry were very deep and subtle. It was not clear that they had anything to do with arithmetic—or the lack of it.

- Arithmetic was in no position to mend the subtle defects of geometry until its own foundations were sound. In particular, a clear concept of *number* was needed.

But the greater the difficulties, the greater the creativity needed to overcome them. The process of reconciliation began with the help of new developments in algebra in the 16th century. This made the methods of arithmetic competitive in geometry for the first time. The process was accelerated by calculus, which gave answers to previously inaccessible questions about lengths and areas of curves. But it was also calculus, with its focus on the "infinitesimal," that most needed a clear concept of number. In 1858, Dedekind realized that calculus, geometry, and the concept of *irrational* could all be clarified in one fell swoop. He defined the concept of *real number* to capture all possible lengths, and thus completed the reconciliation of arithmetic and geometry. The details may be found in Chapter 3.

Exercises

In geometric algebra, a product of three lengths was interpreted as a volume, and there was no interpretation of products of four or more lengths.

2.6.1. With these definitions of addition and multiplication, show that the associative, commutative and distributive laws (Section 1.4) are valid.

2.6.2. Show that the formula $(a + b)^2 = a^2 + 2ab + b^2$ has a natural interpretation in terms of addition and multiplication of lengths.

2.6.3. Also give a geometric interpretation of the identity $b^2 - a^2 = (b - a)(b + a)$.

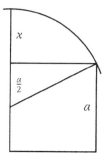

FIGURE 2.20   Geometric solution of a quadratic equation.

In the *Elements*, Book II, Proposition 11, Euclid solves the equation $x^2 = (a - x)a$ by the construction shown in Figure 2.20.

2.6.4. Use Pythagoras' theorem and algebra to check that this is a correct solution.

2.7   Volume

The theory of volume looks much the same as the theory of area at first. The unit of volume is the cube with sides of unit length, and this leads easily to the volume formula for a *cuboid*, the figure whose faces are rectangles. For a cuboid with integer sides we see immediately that volume = width × height × depth by cutting the cuboid into unit cubes. The same formula follows for a cuboid with rational sides by cutting into equal fractional cubes, as we did for rectangles in Section 2.5. Finally, the formula is true for irrational sides either by definition of the product of three lengths (as the Greeks would have it), or by definition of the product of irrational numbers (as we prefer today).

From the cuboid, we can obtain certain other volumes by cutting and pasting, for example, the volume of the *parallelepiped* and the *triangular prism.* The parallelepiped (pronounced "parallel epi ped" where "epi" rhymes with "peppy"), is the three-dimensional figure analogous to the parallelogram, and the triangular prism is obtained by cutting a parallelepiped in half (Figure 2.21).

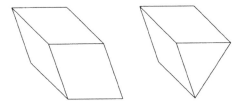

FIGURE 2.21   Parallelepiped and triangular prism.

If we cut a prism off the left end of a cuboid and paste it to the right end (Figure 2.22), we obtain a parallelepiped with a rectangular base and rectangular ends, and volume equal to that of the cuboid, namely, base area × height.

By similarly cutting a prism off the front and pasting it to the back, we obtain a parallelepiped with only the top and bottom rectangular, but still with volume equal to base area × height. Finally, one more cut and paste gives the general parallelepiped, whose faces are arbitrary parallelograms and whose volume is still base area × height. The same is true of a general triangular prism, because it is obtained by cutting a parallelepiped in half.

So far, so good, but these are not typical three-dimensional figures. They are figures of constant cross section, and all we have done so far is operate within their cross sections the way we did in the plane with parallellograms and triangles.

What we really need to know is the volume of a *tetrahedron*, the three-dimensional counterpart of the triangle, because all polyhedra can be built from tetrahedra. The Greeks were unable to cut and paste the tetrahedron into a cuboid, but they found its volume by various ingenious infinite constructions. Perhaps the most elegant is

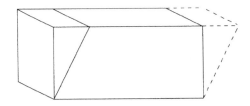

FIGURE 2.22   Volume of a parallelepiped.

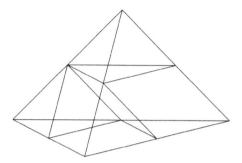

FIGURE 2.23   Pieces of a tetrahedron.

the following, which comes from Euclid. He fills up the tetrahedron with infinitely many prisms.

Figure 2.23 shows a tetrahedron cut into two smaller tetrahedra, the same shape as the original but half its height, and two triangular prisms of equal volume. Each prism has volume $\frac{1}{8}$base × height of the tetrahedron, so their combined volume is $\frac{1}{4}$base × height. Now if each half-size tetrahedron is cut in a similar way, we get half-size prisms; hence each of them is $\frac{1}{8}$ the volume of each original prism. Because there are four half-size prisms, their combined volume is $\frac{1}{4}$ the combined volume of the original two. Continuing this process inside the quarter-size tetrahedra, we get eight quarter-size prisms, with combined volume $\frac{1}{16}$ the combined volume of the original two, and so on.

The total volume of the prisms is therefore

$$\left(\frac{1}{4} + \frac{1}{4^2} + \frac{1}{4^3} + \cdots\right) \text{base} \times \text{height} = \frac{1}{3}\text{base} \times \text{height},$$

by the formula for the sum of the geometric series. But the prisms exhaust the volume of the tetrahedron – they include all points inside the faces of the tetrahedron because the size of the little tetrahedra shrinks toward zero, and hence the volume of the tetrahedron itself is $\frac{1}{3}$base × height.

With such a simple result, it is all the more mysterious that we cannot derive it by cutting and pasting finitely often, as with the area of a triangle. But this is really the case; it can be proved that it is impossible to convert a regular tetrahedron into a cube by cutting it into a finite number of polyhedral pieces. This remarkable result, which was not discovered until 1900, will be proved in Chapter 5.

Exercises

Some of the claims about volumes in the dissection of the tetrahedron should perhaps be checked more carefully.

2.7.1. Explain why the two prisms in Figure 2.23 have equal volume, and why the volume of each is $\frac{1}{8}$base × height of the tetrahedron.

2.7.2. Show that the half-size, quarter-size tetrahedra, ... all lie against the leftmost edge of the tetrahedron, and hence justify the claim that the prisms fill the inside of the tetrahedron.

After the cube and the regular tetrahedron, the next simplest polyhedron is the regular *octahedron*, which is bounded by eight equilateral triangles.

2.7.3. Sketch a regular octahedron, and show that pasting regular tetrahedra on two of its opposite faces gives a parallelepiped.

2.7.4. If one of the triangular faces of the octahedron is taken as the "base" and the distance to the opposite face as the "height," deduce from Exercise 2.7.3 that
$$\text{volume of octahedron} = \frac{4}{3}\,\text{base} \times \text{height.}$$

2.7.5. Also deduce from Exercise 2.7.3 that space may be filled with a mixture of regular tetrahedra and octahedra.

Another way to see the space-filling property, though probably harder, is to prove the following result.

2.7.6.* Show that both the regular tetrahedron and the regular octahedron may be cut into half-sized regular tetrahedra and octahedra.

2.8*   The Whole and the Part

There is an unconscious assumption in cutting and pasting, which we made by speaking of "the" area of a polygon. We are assuming that area is *conserved* in some sense; that if we repeatedly cut and paste, we never get a polygon larger or smaller than the one we started with. This is a blatantly physical assumption, like conservation of mass, and a conscientious geometer would avoid it if possible. To help decide whether we can, let us analyze the process of cutting and pasting more closely.

It is easiest to see the difficulty if we continue to assume that lengths and areas are numbers. There is more than one way to cut a polygon into triangles; what if different ways lead to different numbers? In fact, there is an even more alarming possibility: what if the area $\frac{1}{2}$base × height of a triangle depends on which side we choose to be the base?

The latter possibility can be ruled out, with some difficulty, by Euclid's theory of similar triangles. Triangles are called *similar* if they have the same angles, and Euclid proved that the corresponding sides of similar triangles are proportional. He might also have taken this as an axiom, because it is similar to his axioms about congruence for triangles.

Anyway, assuming proportionality of similar triangles, we can prove the following.

Constancy of base × height   *In any triangle, the product of any side by the corresponding height is constant.*

Proof   Take a triangle ABC and the perpendiculars AD and BE shown in Figure 2.24. Thus if we take BC as the base, AD is the height, and if we take AC as the base, BE is the height. We wish to prove

$$BC \times AD = AC \times BE.$$

To do this we show that triangles ADC and BEC have equal angles. They have angle C in common, and they have right angles at D and

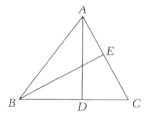

FIGURE 2.24   The constancy of base × height.

E, respectively, hence their remaining angles are also equal because any triangle has angle sum equal to two right angles (Section 2.3).

It follows, by the proportionality of corresponding sides, that

$$\frac{BC}{BE} = \frac{AC}{AD},$$

and therefore

$$BC \times AD = AC \times BE.$$

Thus any two values of base × height for the same triangle are the same. □

The possibility of different areas arising from different subdivisions of the same polygon is more difficult to rule out, and as far as I know this was not done until modern times. For the Greeks, of course, polygons Π and Π' that could be cut and pasted onto each other had equal area *by definition*. Their problem was to show that a polygon could not be cut and pasted onto one that was intuitively "smaller," namely, a *part of itself*. The Greeks could not prove this, and as a last resort Euclid made it one of his assumptions: "the whole is greater than the part."

When areas are numbers, or *numerical areas* as we shall call them, the problem becomes solvable. Hilbert (1899) proved that different subdivisions of the same polygon give the same numerical area. In my opinion, this clinches the case for using numbers in geometry. The main steps in Hilbert's proof are covered in Exercises 2.8.1 to 2.8.3.

Exercises

There are a few preliminaries to Hilbert's proof that can be skipped, because they involve results in the exercises to Section 2.5, and the following result in the same vein: any two subdivisions Σ and Σ' of the same polygon have a *common refinement*, a subdivision Σ'' such that each piece in Σ or Σ' is a union of pieces in Σ''.

The crux of the problem is to prove that any subdivision of triangle Δ into triangles $\Delta_1, \Delta_2, \dots, \Delta_n$ gives the same numerical area. This comes

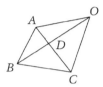

FIGURE 2.25 Areas that the edges span with O.

from proving that *the sum of the $\frac{1}{2}$base × height values for the Δ_k equals the $\frac{1}{2}$base × height value for Δ.*

Hilbert proved this very elegantly using a concept of *signed area* $[ABC]$. $[ABC] = \frac{1}{2}$base × height of triangle ABC when the vertices A, B, and C occur in clockwise order. $[ABC] = -\frac{1}{2}$base × height when A, B, and C occur in counterclockwise order.

2.8.1. Show that $[ABC] = [BCA] = [CAB] = -[ACB] = -[CBA] = -[BAC]$.

The advantage of $[ABC]$ over unsigned area is that it allows the area of any triangle to be expressed as the sum of areas that its edges "span" with a common origin O, as shown in Figure 2.25.

2.8.2. Show that $[OAB] = [OAD] + [ABD]$, $[OBC] = [ODC] + [DBC]$, and hence

$$[ABC] = [OAB] + [OBC] + [OCA].$$

Check that the same result holds for other positions of O outside triangle ABC.

If we have a triangle $\Delta = ABC$ cut into triangles $\Delta_k = A_k B_k C_k$ we use the previous exercise to write $[A_k B_k C_k] = [OA_k B_k] + [OB_k C_k] + [OC_k A_k]$ for each k, expressing the area of Δ_k as the sum of the areas its edges span with O.

2.8.3. Show that if the equations $[A_k B_k C_k] = [OA_k B_k] + [OB_k C_k] + [OC_k A_k]$ are added, the areas spanned by edges inside Δ cancel out. Thus the only terms remaining on the right-hand side are of the form $[OEF]$, where EF is a segment of one of the sides of Δ. Deduce that

$$[A_1 B_1 C_1] + [A_2 B_2 C_2] + \cdots + [A_k B_k C_k] = [ABC],$$

and conclude that any subdivision of Δ gives the same numerical area.

We say that polygons Π and Π' are *equidecomposable* if Π may be cut into polygonal pieces that can be pasted together to form Π'. Two equidecomposable polygons have the same numerical area, because their areas are the sums of the numerical areas of the same pieces.

More surprisingly, any two polygons of equal numerical area are equidecomposable. The result is not hard, but it was not noticed until the 19th century, possibly because the Greek idea that equal area meant equidecomposability *by definition* lingered until then. The proof can be broken down into the following steps:

1. Show that each polygon can be cut into a finite number of triangles.

2. Show that any triangle can be cut and pasted into a rectangle with given base, say 1.

3. Given any polygon Π, cut it into triangles, and cut and paste each triangle into a rectangle of base 1.

4. Stack up the rectangles, obtaining a single rectangle R of base 1, equidecomposable with Π.

5. Do the same with the other polygon Π′, obtaining the same rectangle R (because Π′ has the same area as Π).

6. Conclude, by running the construction from Π to R, and then the construction from Π′ to R in reverse, that Π is equidecomposable with Π′.

After making this breakdown, the only steps that need any work are the first two.

2.8.4. Explain how to do Step 1.

Step 2 is the hardest, and is best broken down into two substeps.

2.8.5. Show that any triangle may be cut and pasted into a rectangle.

2.8.6. Use Figure 2.26 to explain how any rectangle may be cut and pasted into a rectangle of width 1.

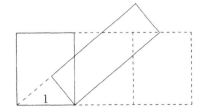

FIGURE 2.26 Equidecomposable rectangles.

2.9   Discussion

The Pythagorean Influence

The Pythagorean philosophy that "all is number" has come down to us through legends rather than hard evidence from Pythagoras' time. Nevertheless, these legends were persistent enough to influence the development of mathematics and physics until the present day. The story goes that the Pythagoreans came to believe in the power of numbers through discovering their role in music. They found, by studying the sounds of plucked strings, that the most harmonious notes were produced by strings whose lengths were in simple integer ratios. Given that the strings are of the same material, thickness, and tension, the most harmonious pairs of notes occur when the ratio of lengths is 2:1 (the octave), 3:2 (the "perfect fifth"), and 4:3 (the "perfect fourth").

Even today one must admit that this is a discovery good enough to build a dream on—if the subjective experience of harmony can be explained by numbers, perhaps *anything* can. Whether or not the Pythagoreans actually thought this, the idea sooner or later caught on and inspired other persistent dreams. The best known was the "harmony of the spheres," which tried to explain the position of the planets by numbers. It haunted astronomy from Aristotle (around 350 B.C.) to Johannes Kepler (1571–1630), until Isaac Newton came up with a better idea, the theory of gravitation.

If the Pythagoreans really believed that "all is number" (meaning natural numbers and their ratios), then it is, of course, ironic that their own philosophy should be brought down by the Pythagorean theorem and the irrationality of $\sqrt{2}$. An even greater irony, however, is that *Pythagorean music theory itself is fundamentally irrational.* This fact comes to light as soon as one tries to compare the "size" of the octave and the perfect fifth. According to another legend, Pythagoras himself tried to do this, finding that the interval of 7 octaves is very close, but not equal, to the interval of 12 perfect fifths. The pitch of a string is lowered by 7 octaves when its length is multiplied by $2^7 = 128$, and by 12 perfect fifths when its length is multiplied by $(3/2)^{12} = 129.746\ldots$.

Similarly, the pitch is lowered by m octaves when length is multiplied by 2^m, and by n perfect fifths when length is multiplied by $(3/2)^n$. *But there are no natural numbers such that*

$$2^m = \left(\frac{3}{2}\right)^n,$$

as this implies

$$2^{m+n} = 3^n,$$

which is absurd because 2^{m+n} is even and 3^n is odd. Thus there are no natural numbers m and n such that m octaves equals n perfect fifths—in other words, the ratio of the octave to the perfect fifth is irrational.

Geometry, Measurement, and Numbers

I have claimed that the discovery of irrational lengths led to the separation of geometry from number theory in Greek mathematics. But perhaps this was only because rigor demanded a separate development of geometry, as long as there was no rigorous definition of irrational number, it was necessary to work with lengths. Rigor and precision are necessary for communication of mathematics to the public, but they are only *last stage* in the mathematician's own thought. New ideas generally emerge from confusion and obscurity, so they cannot be grasped precisely until they have first been grasped vaguely and even inconsistently. We know, for example, that Archimedes discovered results on the area and volume of curved figures by dubious methods, then revised his proofs to make them rigorous. Only the rigorous versions were known until 1906, when Heiberg discovered a lost manuscript revealing Archimedes' original methods (see Heath (1912)). Thus it is quite possible that the Greeks thought about irrational numbers but wrote about lengths for public consumption.

Even if this is so, shouldn't geometry be about lengths? Its name means "land measurement," after all. Well, this probably has more to do with the legendary origins of geometry than its actual use in ancient Greece. Plato believed that geometry was not really about land measurement but about different types of numbers. In the

Epinomis, a work due to Plato or one of his disciples, we find the remarkable statement:

> ... what is called by the very ridiculous name *mensuration* (*geometria*), but is really a manifest assimilation to one another of numbers which are naturally dissimilar, effected by reference to areas. Now to a man who can comprehend this, 'twill be plain that this is no mere feat of human skill, but a miracle of God's contrivance. [From the translation of the *Epinomis* by Taylor (1972), p. 249.]

Even in a profession where measurement is important, the ancients were more impressed with theory. In the 1st century B.C., Vitruvius wrote in his *Ten Books on Architecture,* Introduction to Book IX:

> Pythagoras showed that a right angle can be formed without the contrivances of the artisan. Thus the result which carpenters reach very laboriously, but scarcely to exactness, with their squares, can be demonstrated to perfection from the reasoning and methods of his teaching.

Presumably he was thinking of the converse Pythagorean theorem, according to which lengths a, b, and c for which $a^2 + b^2 = c^2$ make a triangle with a right angle between the sides a and b. (This follows easily from Pythagoras' theorem itself, and the side-side-side congruence axiom. Namely, construct a right angle with sides a and b, so the hypotenuse joining these sides has length c by Pythagoras' theorem. Then we have a right-angled triangle with sides a, b, and c, and it is the *only* triangle with these sides, by the congruence axiom.) He then described the construction of a right angle by combining rods of lengths 3, 4, and 5 in a triangle, and finally he suggested an application:

> This theorem affords a useful means of measuring many things, and it is particularly serviceable in the building of staircases in buildings, so that steps may be at their proper levels.

It has often been claimed that the $(3, 4, 5)$ triangle was used in ancient times to construct right angles, but this is the oldest reference to it that I know.

FIGURE 2.27 Geometric algebra on a Greek coin.

Just as we cannot be sure how the Greeks viewed irrationals, we cannot tell how they viewed the so-called *geometric algebra* in Euclid's *Elements*. One would not expect a full understanding of algebra, especially not with the Greeks' unsuitable notation. But they may have caught a glimpse of it. The geometric interpretation of $(a + b)^2 = a^2 + 2ab + b^2$ (Exercise 2.6.2) was so well known in ancient Greece that the figure actually appeared on coins. Figure 2.27 shows a photograph of one; this photograph was given to me by Benno Artmann.

The diagonal line in the figure is probably a construction line; the corresponding figure in Euclid (Book II, Proposition 4) uses this line (but drawn all the way across) to divide the vertical line into the same two parts, a and b, as the horizontal line.

3

CHAPTER

Coordinates

3.1 Lines and Circles

The most important step in geometry since ancient times was the introduction of coordinates by Descartes in his *Geometry* of 1637. Coordinates are a simple idea, but not much use without algebra and a symbolic notation, which is probably why the idea did not take off earlier. The time was ripe for coordinates in 1637, because algebra and its notation had matured over the preceding century, to a level similar to high school algebra today. In fact, Fermat in 1629 hit on the same idea as Descartes, and he illustrated it with similar results, but they were not published at the time.

The coordinates of a point P of the plane are its distances (x, y) from perpendicular axes OY and OX (Figure 3.1). O is called the *origin* of coordinates, and its own coordinates are $(0, 0)$.

Distances are normally regarded as positive numbers, so this scheme initially applies only to points P above and to the right of O, the so-called *positive quadrant*. In fact, Fermat and Descartes did not look beyond this quadrant, because they did not consider the possibility of negative distances. But if points to the left of O are

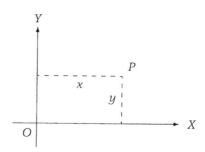

FIGURE 3.1   Coordinates of a point.

given negative x coordinates and points below O are given negative y coordinates, everything works smoothly. (This is one of the benefits of making the rules of arithmetic for negative quantities the same as the rules for positive quantities, as we did in Section 1.4.) In particular, some curves that look "sawn off" in the positive quadrant have natural extensions to the rest of the plane.

For example, a straight line through the origin (other than OY) has the property that the ratio y/x of distances from the axes is a constant, say m. This gives us the equation $y = mx$, and the negative values of x and y satisfying the same equation lie on the same straight line (Figure 3.2).

Generalizing this idea slightly, we arrive at the general equation $ax + by = c$, which gives any straight line by suitable choice of constants a, b, and c. This is why we call this equation *linear*.

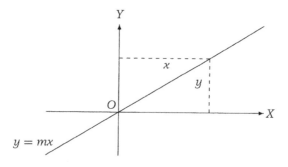

FIGURE 3.2   Straight line through the origin.

The next most important curve is the circle, whose points have the property of being at constant distance from its center. If we take the center to be O, and radius r, then Pythagoras' theorem tells us that

$$x^2 + y^2 = r^2,$$

because r is the hypotenuse of a right-angled triangle with sides x and y. Similarly, the points at distance r from an arbitrary point (a, b) satisfy the equation

$$(x - a)^2 + (y - b)^2 = r^2,$$

hence this is the equation for a general circle.

These equations are no doubt very familiar to most readers, but it is worth reflecting on what we assumed in deriving them. Among other things, we used the existence of rectangles, similar triangles of different sizes, and Pythagoras' theorem. In Chapter 2 we saw that all of these are characteristic of Euclidean geometry, each of them equivalent to the parallel axiom.

The representation of straight lines and circles by equations in fact gives yet another way to define Euclidean geometry. This did not dawn on Fermat and Descartes—they took their geometry straight out of Euclid and saw the coordinates merely as a way of handling it more efficiently—but it became important when non-Euclidean geometry was discovered in the 19th century. One finally had to wonder: what *is* Euclidean geometry? And what is non-Euclidean geometry, for that matter? The beauty of coordinates is that they allow all geometries to be built on a common foundation of numbers, with different geometries distinguished by different equations. When we look at geometry this way, Euclidean geometry turns out to be the one with the simplest equations.

Of course, the improved handling obtained by the use of coordinates is also important. Coordinate geometry is called *analytic* because situations are "analyzed" rather than "synthesized." Typically, points, lines, and circles are found by solving equations, rather than by ruler and compass construction as in Euclid. Thanks to the power of algebra, analysis is a method of much greater scope. As we shall see, it can even show the *impossibility* of certain constructions sought by the Greeks.

Exercises

Some of Euclid's axioms correspond to familiar facts about equations, for example, the axiom that a (unique) line can be drawn through any two points.

3.1.1. Find the (unique) linear equation satisfied by two points (a_1, b_1) and (a_2, b_2).

Also prove the parallel axiom.

3.1.2. How would you recognize a line parallel to the line $ax + by = c$ from its equation? Show that its equation has no solution in common with $ax + by = c$ and that there is only one such line through a given point outside the line $ax + by = c$.

3.2 Intersections

The difference between analytic and synthetic geometry can be illustrated with Euclid's very first proposition, the construction of the equilateral triangle on a line segment AB. Recall from Section 2.2 that Euclid did this by finding the intersection of two circles, one with center A and radius AB and the other with center B and radius BA. It follows that each point of intersection is distant from both A and B by the length of AB and hence forms an equilateral triangle. Analytic geometry takes its cue from this construction, but it also finds the intersections by finding the common solutions of the equations to the two circles.

For example, if $A = (-1, 0)$ and $B = (1, 0)$ the two circles have radius 2 and hence their equations are

$$(x + 1)^2 + y^2 = 2^2,$$
$$(x - 1)^2 + y^2 = 2^2.$$

Subtracting the second equation from the first leads to $x = 0$ (as you would expect), and substituting this back in the first equation gives

$$1 + y^2 = 4,$$

whence $y = \pm\sqrt{3}$. Thus the points of intersection are $(0, \pm\sqrt{3})$, either of which can be taken as the third vertex of the equilateral triangle.

Euclid's argument is short and sweet, but it has one defect. It does not follow from his axioms! His axioms guarantee only the existence of circles, not their intersections. This defect can be repaired by introducing axioms about intersections, but only with difficulty, because it is hard to foresee all the situations that may arise. The great advantage of coordinates is that all questions about intersections become questions about solutions of equations, which algebra can answer. In this case, the algebra shows that existence of the intersection depends on existence of the number $\sqrt{3}$.

In fact, we have the following theorem.

Nature of constructible points   *Points constructible by ruler and compass have coordinates obtainable from 1 by the rational operations* $+, -, \times, \div,$ *and square roots.*

Proof   Recall from Section 2.2 that the given constructions are:

- To draw a straight line between any two given points.

- To draw a circle with given center and radius.

In the beginning, we are given only the unit of length, which we may take to be the line segment between $(0, 0)$ and $(1, 0)$. All points are constructed as intersections of lines and circles, so it will suffice to show the following:

1. The line through (a_1, b_1) and (a_2, b_2) has an equation with coefficients obtainable from a_1, a_2, b_1, and b_2 by rational operations.

2. The circle with center (a, b) and radius r has an equation with coefficients obtainable from a, b, and r by rational operations.

3. The intersection of two lines has coordinates obtainable from the coefficients of their equations by rational operations.

4. The intersection of a line and a circle has coordinates obtainable from the coefficients of their equations by rational operations and square roots.

5. The intersection of two circles has coordinates obtainable from the coefficients of their equations by rational operations and square roots.

These facts are confirmed by calculations like those we have already considered.

1. The line through (a_1, b_1) and (a_2, b_2) has equation

$$(b_1 - b_2)x - (a_1 - a_2)y = a_2 b_1 - a_1 b_2,$$

as may be checked by substituting the points $x = a_1$, $y = b_1$ and $x = a_2$, $y = b_2$. All the coefficients come from a_1, a_2, b_1, b_2 by rational operations.

2. The circle with center (a, b) and radius r has equation

$$(x - a)^2 + (y - b)^2 = r^2,$$

as we already know.

3. The intersection of two lines

$$a_1 x + b_1 y = c_1 \quad \text{and} \quad a_2 x + b_2 y = c_2$$

is computed from a_1, b_1, c_1 and a_2, b_2, c_2 by rational operations. Just recall the usual process for solving a pair of linear equations.

4. The intersection of the line

$$a_1 x + b_1 y = c$$

with the circle

$$(x - a_2)^2 + (y - b_2)^2 = r^2$$

is found by substituting $x = (c - b_1 y)/a_1$, from the equation of the line, in the equation of the circle (unless $a_1 = 0$, in which case we substitute the similar expression for y). This gives a quadratic equation for x, the coefficients of which are rational in the co-efficients of the line and the circle. The quadratic formula gives x from the new coefficients by further rational operations and (possibly) a square root. Finally, by substituting x back in the equation of the line, we obtain y by further rational operations.

5. The intersection of the two circles

$$(x - a_1)^2 + (y - b_1)^2 = r_1^2 \quad \text{and} \quad (x - a_2)^2 + (y - b_2)^2 = r_2^2$$

is found by expanding these equations to

$$x^2 + 2a_1x + a_1^2 + y^2 + 2b_1y + b_1^2 = r_1^2,$$
$$x^2 + 2a_2x + a_2^2 + y^2 + 2b_2y + b_2^2 = r_2^2,$$

and subtracting the second from the first to obtain the linear equation

$$2(a_1 - a_2)x + 2(b_1 - b_2)y = r_1^2 - r_2^2 + a_2^2 - a_1^2 + b_2^2 - b_1^2.$$

All the coefficients are rational combinations of the original coefficients, so we are now reduced to the situation just considered, the intersection of a circle and a line. It follows that the coordinates of the intersections may be found by rational operations and square roots. □

It follows from this theorem that if a number is *not* expressible by rational operations and square roots, the corresponding point is not constructible by ruler and compass. This opens the way for algebraic attack on problems of constructibility, and in this way some of the Greek problems were shown to be unsolvable, after 2000 years of unsuccessful attempts to solve them. The simplest example of a nonconstructible number is $\sqrt[3]{2}$; see the exercises.

Exercises

The theorem on the nature of constructible points also has a converse: *there is a ruler and compass construction of any point with coordinates obtainable by rational operations and square roots (of positive numbers).*

3.2.1. Explain how to do addition and subtraction.

The keys to multiplication and division are the similar triangles in Figure 3.3.

3.2.2. Explain why the lengths are as shown in Figure 3.3.

Likewise, similar right-angled triangles are the key to constructing a square root (Figure 3.4).

3.2.3. Explain why the lengths are as shown in Figure 3.4.

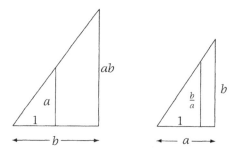

FIGURE 3.3   Constructing the product and quotient of lengths.

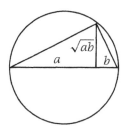

FIGURE 3.4   Constructing the square root of a length.

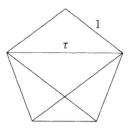

FIGURE 3.5   Regular pentagon.

It follows from this converse theorem that any length expressible by rational operations and square roots is constructible by ruler and compass. This gives an easy way to see that certain figures are constructible. For example, we can say immediately that the length $\tau = \frac{1+\sqrt{5}}{2}$ is constructible. This shows that the regular pentagon is constructible, because the regular pentagon with unit sides has diagonal τ (Figure 3.5).

3.2.4.* Prove that the diagonal is τ. (It may help to use the lines shown in Figure 3.5 and find some similar triangles.)

Now let us see why the number $\sqrt[3]{2}$ is *not* constructible. An elementary proof was discovered by the number theorist Edmund Landau (1877–1938) when he was still a student. He starts with the set F_0 of rational numbers and considers successively larger sets $F_1, F_2, \ldots$, each obtained from the one before by adding the square root of one of its members, and then applying rational operations. The aim is to show that $\sqrt[3]{2}$ is never reached in this way, because its presence would yield a contradiction.

3.2.5.* Let F_0 be the set of rational numbers and let $F_{k+1} = \{a + b\sqrt{c_k} : a, b \in F_k\}$ for some $c_k \in F_k$. Show that the sum, difference, product, and quotient of any members of F_{k+1} are also members. (Hence each F_k is a *field*, as defined in the exercises in Section 1.4.)

3.2.6.* Show that if $a, b, c \in F_k$ but $\sqrt{c} \notin F_k$ then $a + b\sqrt{c} = 0 \Leftrightarrow a = b = 0$.

3.2.7.* Suppose $\sqrt[3]{2} = a + b\sqrt{c}$ where $a, b, c \in F_k$, but that $\sqrt[3]{2} \notin F_k$. (We know $\sqrt[3]{2} \notin F_0$ because $\sqrt[3]{2}$ is irrational by Exercise 1.6.8.) Cube both sides and deduce that

$$2 = a^3 + 3ab^2 c \quad \text{and} \quad 0 = 3a^2 b + b^3 c.$$

3.2.8.* Deduce from Exercise 3.2.7* that $a - b\sqrt{c} = \sqrt[3]{2}$ also, which is a contradiction.

3.3 The Real Numbers

The results of the last section throw new light on the relationship between numbers and geometry. We knew from the beginning that the rational numbers cannot represent all lengths, because irrational square roots occur in even the simplest figures. Now we know that all lengths actually needed in the geometry of straight lines and circles arise from rational operations and square roots. If we want to treat these lengths as numbers, therefore, it suffices to understand square roots.

Nevertheless, this is a hard problem. Not only do we have to understand $\sqrt{2}, \sqrt{3}$, and so on, but also more complicated numbers such as $\sqrt{1 + \sqrt{2}}$ and $\sqrt{\sqrt{2} + \sqrt{3}}$, because the corresponding lengths are all constructible by Exercises 3.2.1 to 3.2.3. The simplest solution was discovered by Dedekind in 1858, when he decided that a better understanding of *all* irrational numbers was desirable. It comes from reflecting on the "position" an irrational number occupies among the rationals.

Consider $\sqrt{2}$. It is less than each of the numbers

$$2$$
$$1.5$$
$$1.42$$
$$1.415$$
$$1.4143$$
$$1.41422$$
$$1.414214$$
$$1.4142136$$
$$\ldots$$

because each of these numbers has a square greater than 2. Likewise, it is greater than each of

$$\ldots$$
$$1.4142135$$
$$1.414213$$
$$1.41421$$
$$1.4142$$
$$1.414$$
$$1.41$$
$$1.4$$
$$1$$

because each of the latter numbers has square less than 2.

These two lists of rational numbers give a rough idea of the position of $\sqrt{2}$ among all the rationals. Its exact position can be specified by the set L of *all* the positive rationals with squares < 2 (L is for "lower"), and the set U of positive rationals with squares > 2 (U is for "upper"), because there is no other number that fits between these two sets. Thus $\sqrt{2}$ can be recognized in its absence, as it were, by the two sets into which the positive rationals separate. It dawned on Dedekind that this is all we need to know about $\sqrt{2}$; it may as well be *defined* as the pair of sets of rationals (L, U).

It takes a while to get your breath back after first seeing this idea, because everyone thinks that $\sqrt{2}$ is already "there," and we only have to compute it. But no, the real problem is to *define* it, and hence to know what it is we are computing. The beauty of Dedekind's definition is that it requires nothing new, only sets of objects already assumed to exist: the rational numbers. Since we know that $\sqrt{2}$ is not itself a rational number, this is as simple as the definition can be.

The general idea is to imagine that the rationals are separated by an irrational number, then take the separation—or *cut* as Dedekind called it—to *be* the number. This enables us to define all positive irrational numbers in one fell swoop as follows.

Definition A *positive irrational number* is a pair (L, U) of sets of positive rationals such that

- L and U together include all positive rationals.

- Each member of L is less than every member of U.

- L has no greatest member and U has no least member.

It is not necessary to stick to positive irrationals either, because the sets L and U can be taken from the set of all rationals. However, it is convenient to keep the restriction to positive rationals a little longer, as it makes it easier to define addition and (especially) multiplication.

Each rational number a also makes a cut in the set of rationals, of course, into the set $L_a = \{$rationals $< a\}$ and the set $U_a = \{$rationals $\geq a\}$. The only difference is that U_a has a least member, namely a. This prompts us to define positive *real* numbers so as to include both rationals and irrationals, by weakening the third condition in the definition of irrational number to say only that L has no greatest member.

Definition A *positive real number* is a pair (L, U) of sets of positive rationals such that

- L and U together include all positive rationals.

- Each member of L is less than every member of U.
- L has no greatest member.

Then we are ready to define addition and multiplication for all positive real numbers (or *reals*, for short).

Definition   If (L_1, U_1) and (L_2, U_2) are *positive reals* then

- Their sum is the real number (L, U) such that

$$L = \{x_1 + x_2 : x_1 \in L_1 \text{ and } x_2 \in L_2\}$$

and U consists of the remaining positive rationals.

- Their product is the real number (L, U) such that

$$L = \{x_1 x_2 : x_1 \in L_1 \text{ and } x_2 \in L_2\}$$

and U consists of the remaining positive rationals.

After this, $+$ and $\times$ can be extended to negative reals the same way they were for negative integers in Section 1.4. The set of all real numbers is denoted by $\mathbb{R}$.

To see that there is method in this madness, let us check that $\sqrt{2}\sqrt{2} = 2$.

By definition of $\sqrt{2}$ and the definition of multiplication, the L for $\sqrt{2}\sqrt{2}$ is $\{x_1 x_2 : x_1^2 < 2 \text{ and } x_2^2 < 2\}$, where the x_1 and x_2 are rational.

It follows that each $x_1^2 x_2^2$ is less than $2 \times 2 = 4$, and therefore $x_1 x_2$ is a rational x less than 2.

Conversely, any rational x less than 2 can be written as $x = x_1 x_2$, where x_1 and x_2 are rationals with $x_1^2 < 2$ and $x_2^2 < 2$. This is because the rationals crowd together arbitrarily closely, and hence so do their squares. It follows that there are rational squares as close as we please to x, and if x_1 is chosen with x_1^2 sufficiently close to x and $x_2 = x/x_1$, then $x_1 x_2 = x$ and both $x_1^2 < 2$ and $x_2^2 < 2$.

Thus the L for $\sqrt{2}\sqrt{2}$ is $\{x < 2\}$, which is the L for 2, as required.

Exercises

It is now possible to appreciate Dedekind's claim that

in this way we arrive at real proofs of theorems (as, e.g., $\sqrt{2}\sqrt{3} = \sqrt{6}$), which to the best of my knowledge have never been established before. [Dedekind (1872), p.22]

As we can see from the example $\sqrt{2}\sqrt{2} = 2$, proving such equations for numbers is very different from proving them for lengths, mainly because the product of irrational numbers is defined so differently from the product of lengths. Recall from Section 2.5 that the Greeks defined the product of lengths $\sqrt{2} \times \sqrt{3}$ to be the rectangle with sides $\sqrt{2}$ and $\sqrt{3}$, and it could be shown equal to $\sqrt{6}$ only by cutting and pasting to form a rectangle with sides $\sqrt{6}$ and 1. Dedekind's theory of irrational numbers gives us a rigorous alternative.

3.3.1. Prove that the numbers $\sqrt{2}$, $\sqrt{3}$, and $\sqrt{6}$ satisfy $\sqrt{2}\sqrt{3} = \sqrt{6}$.

I admit this proof is tedious, but once one such proof has been done, the same routine can be followed in other cases, like the following.

3.3.2. Prove that the numbers $\sqrt[3]{2}$, $\sqrt[3]{3}$, and $\sqrt[3]{6}$ satisfy $\sqrt[3]{2}\sqrt[3]{3} = \sqrt[3]{6}$.

The corresponding theorem about lengths cannot be proved geometrically, because the lengths are not constructible! Thus Dedekind's definition of product of numbers gives us everything we could previously do with the product of lengths, and more. It is not only *possible* to treat lengths as numbers, but it is an advantage.

While on the subject of defining irrational numbers, it should be explained where infinite decimals like $\sqrt{2} = 1.41421356\ldots$ fit in. As the arrangement of numbers $> \sqrt{2}$ and numbers $< \sqrt{2}$, on page 78 suggests, the symbol $1.41421356\ldots$ is a concise way to describe the infinite sequence of rationals $1, 1.4, 1.414, 1.4142, 1.41421, \ldots$, which in turn is part of the lower set L for $\sqrt{2}$. The sequence is said to be *cofinal* with L, because they "end at the same place"; L consists of the rationals less than members of the sequence. For this reason, $1.41421356\ldots$ contains the same information as L, and hence can also serve to represent $\sqrt{2}$.

The main advantage of $1.41421356\ldots$ is that we understand its finite decimal approximations $1, 1.4, 1.414, 1.4142, 1.41421, \ldots$ and we are used to computing with them. However, it is not as easy to define sum and product for infinite decimals as it is for Dedekind cuts.

3.3.3. Try to define sum and product for infinite decimals.

Apart from this, the main disadvantage of $1.41421356\ldots$ is the lack of any apparent pattern in the sequence of its digits. In fact, the simplest way to describe its finite decimal approximations is to say that they are respectively the largest 1-digit, 2-digit, 3-digit, ... decimals whose squares are less than 2. Thus we end up essentially repeating Dedekind's definition.

There is in fact another way to describe $\sqrt{2}$ by a process with an infinite, but repeating, pattern. This is the *continued fraction algorithm*, which will be described in Chapter 8. It is definitely not the case that the infinite decimal for $\sqrt{2}$ eventually becomes repeating, because this does not happen for any irrational number.

3.3.4. Let $x = 0.\overline{a_1 a_2 \ldots a_k}$ be a number whose infinite decimal consists of the sequence $a_1 a_2 \ldots a_k$ repeated indefinitely. Using the infinite geometric series, or otherwise, show that x is rational.

3.3.5. Let $y = 0.b_1 b_2 \ldots b_j \overline{a_1 a_2 \ldots a_k}$ be a number whose infinite decimal, after the first j places, consists of the sequence $a_1 a_2 \ldots a_k$ repeated indefinitely. Show that y is also rational. (Such a decimal is called *ultimately periodic*.)

3.3.6. Show that any rational number has an ultimately periodic decimal.

3.4   The Line

Having seen how individual lengths, like $\sqrt{2}$, can be reborn as numbers, the next step is to see whether these numbers make up anything we would recognize as a line.

One crucial property they have is *order*: if α and β are any distinct real numbers, then either $\alpha \leq \beta$ or $\beta \leq \alpha$. In fact, if $\alpha = (L_\alpha, U_\alpha)$ and $\beta = (L_\beta, U_\beta)$, it is natural to say that $\alpha \leq \beta$ if and only if L_α is *contained in L_β*, because this captures the idea that α separates the rationals at a position $\leq$ the position where they are separated by β. If L_α is not contained in L_β then there is a rational r in L_α but not in L_β, in which case all members of L_β are less than r. Then L_β is contained in L_α, and hence $\beta \leq \alpha$ by our definition. Thus the real numbers have an order, like points on a line.

The second crucial property of the line is what Dedekind called its *continuity*, or absence of gaps. Do the real numbers have this property? Weil, the real numbers were created precisely by filling all the gaps in the rationals. A gap occurs where the rationals split into a lower set L with no greatest member and an upper set U with no least member, and we filled each such gap by the irrational number (L, U). We could even say that the number (L, U) is the gap in the rationals!

Thus the irrationals fill all gaps in the set $\mathbb{Q}$ of rationals, by definition. Can the resulting set $\mathbb{R}$ of reals have gaps? The answer is *no*, because a gap in $\mathbb{R}$ implies an "unfilled gap" in $\mathbb{Q}$. In fact, if $\mathbb{R}$ is separated into a lower set $\mathcal{L}$ and an upper set $\mathcal{U}$, consider the following sets of rationals r:

$$L = \{r : r \le \text{ some member of } \mathcal{L}\},$$
$$U = \{r : r \ge \text{ some member of } \mathcal{U}\}.$$

Because $\mathcal{L}$ and $\mathcal{U}$ together include all reals, L and U together include all rationals. And because $\mathcal{L}$ and $\mathcal{U}$ have no members in common, neither do L and U. L and U therefore define a number (L, U). But then (L, U) is either the least member of $\mathcal{U}$ or the greatest member of $\mathcal{L}$, so there is no gap where $\mathbb{R}$ is separated.

The "no gaps" property of $\mathbb{R}$ is now called *completeness*, because Dedekind's word "continuity" is used for a related property of functions or curves. We also say that $\mathbb{R}$ is the *completion of* the set $\mathbb{Q}$ of rationals. At any rate, ordering and completeness are exactly what we were looking for to model the concept of line in geometry, so $\mathbb{R}$ fits the bill. We often call $\mathbb{R}$ the *real line*. It now remains to check that pairs (x, y) of real numbers can be made to behave like points of the plane, and the conversion of geometry to arithmetic will be complete. We shall do this in the next section.

Identifying the line with the real numbers has other advantages, apart from allowing the free use of arithmetic in geometry. It gives answers to questions that cannot really be settled by geometric intuition, because they involve the "infinitely small." For example, most people have the feeling, at first, that $0.999999999\ldots$ cannot be equal to 1, because it seems to be less than 1 by an "infinitesimal amount"; maybe 1 is the "next number" after $0.999999999\ldots$. Such feelings are dispelled by Dedekind's picture of real numbers. In fact, we can say definitely that:

1. There is no such thing as the "next point," because there is no such thing as the "next real number." If α and β are distinct real

numbers then $(\alpha + \beta)/2$ is a number that lies strictly between them.

In fact, there is a rational number strictly between them. For example, if $\alpha < \beta$, take any number in the lower set for β that is not in the lower set for α. (Here it is convenient that we defined reals so that their lower sets never have greatest members.)

2. There are no "infinitesimal distances" between points, that is, distances that are nonzero yet less than any positive rational. This is because there is no positive number less than all positive rationals. In fact, if α is a positive real number, then the lower set for α must include a positive rational, and all numbers in the lower set are less than α.

Exercises

Another important property of $\mathbb{R}$ is the existence of *least upper bounds*: if the numbers in some set S are all $\leq$ some number α, then there is a least number $\lambda \geq$ all members of S. This number $\lambda \leq \alpha$ is called the *least upper bound of S*.

We can obtain λ by taking its lower set to be the union of all the lower sets L_β of members β of S. (That is, L_λ is the set of all the members of all the sets L_β.)

3.4.1. Deduce from this definition that $\beta \leq \lambda$ for each β in S.

3.4.2. Show also that if $\mu < \lambda$ then $\mu <$ some β in S.

The existence of least upper bounds is in fact another way to state the completeness of $\mathbb{R}$.

3.4.3. Suppose $\mathbb{R}$ is separated into a lower set $\mathcal{L}$ and an upper set $\mathcal{U}$. Use the existence of least upper bounds to show this separation is not a gap. (That is, either $\mathcal{L}$ has a greatest member or $\mathcal{U}$ has a least member.)

3.4.4. Conversely, use the nonexistence of gaps to find a least upper bound for any bounded set S.

The "no infinitesimals" property of $\mathbb{R}$ can be stated in another way that goes back to Archimedes. It is called the *Archimedean axiom*; it says that if α and β are positive numbers, with $\alpha < \beta$, then there is a natural number n with $n\alpha > \beta$.

3.4.5. Prove the Archimedean axiom.

3.5 The Euclidean Plane

Now that we have the line, as the set of real numbers x, the plane is obtained by a simple trick. It is the set of *ordered pairs* of real numbers, (x, y). In honor of Descartes, this set is called the *cartesian product*, $\mathbb{R} \times \mathbb{R}$, of the set $\mathbb{R}$ of reals with itself. The main difference between Descartes and us is that he supposed the plane to exist, then gave each point in it a coordinate pair (x, y); we suppose only that numbers exist, we say the coordinate pair (x, y) *is* a point, and that the set of these points is the plane.

We also have to define the distance between points, which is not hard, because we know what it should be from previous experience.

Definition The *Euclidean distance* between $P_1 = (x_1, y_2)$ and $P_2(x_2, y_2)$ is

$$d(P_1, P_2) = \sqrt{(x_2 - x_1)^2 + (y_2 - y_1)^2}.$$

This is prompted by the Pythagorean theorem, because we expect the line segment from (x_1, y_1) to (x_2, y_2) to be the hypotenuse of a right-angled triangle with sides $x_2 - x_1$ and $y_2 - y_1$ (Figure 3.6).

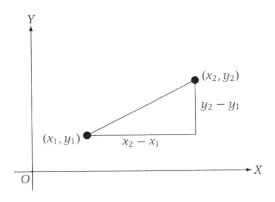

FIGURE 3.6 The distance-defining triangle.

The set $\mathbb{R} \times \mathbb{R}$ with this distance function is called the *Euclidean plane*. As we know, the Pythagorean theorem is a characteristic statement of Euclid's geometry, and by defining distance as we did we have made the theorem true *by definition* in $\mathbb{R} \times \mathbb{R}$. With a different choice of distance function we can get a *non*-Euclidean plane, as we shall see in Section 3.8*.

A *line* is defined to be the set of points (x, y) satisfying an equation of the form $ax + by = c$. A *circle* is defined to be the set of points (x, y) at constant distance r from a point (a, b). It follows from the definition of distance that the equation of the circle is $(x-a)^2 + (y-b)^2 = r^2$, as expected. Thus we can re-create the basic concepts of Euclid's geometry in terms of numbers, with the added advantage that Euclid's unstated assumptions about the existence of intersections are guaranteed. There are enough real numbers to solve all the equations that arise when we seek intersections of lines and circles.

Basic properties of distance *It follows from the definition of distance that*

1. *The set of points equidistant from two distinct points is a line.*

2. *Any line is the equidistant set of two points.*

3. *Each point of the plane is determined by its distances from three points not in a line.*

Proof

1. If $P_1 = (x_1, y_1)$ and $P_2 = (x_2, y_2)$ are any two points, then the points (x, y) equidistant from them both satisfy

$$(x - x_1)^2 + (y - y_1)^2 = (x - x_2)^2 + (y - y_2)^2,$$

which is equivalent to the equation of a line, namely,

$$2(x_2 - x_1)x + 2(y_2 - y_1)y = x_2^2 - x_1^2 + y_2^2 - y_1^2.$$

2. The latter equation represents an arbitrary line $ax + by = c$, provided we can find some constant k such that

$$2(x_2 - x_1) = ka$$
$$2(y_2 - y_1) = kb$$
$$x_2^2 - x_1^2 + y_2^2 - y_1^2 = kc.$$

Substituting $x_2 = x_1 + \frac{k}{2}a$ from the first equation, and $y_2 = y_1 + \frac{k}{2}b$ from the second in the third gives an equation from which we find

$$k = \frac{4(c - ax_1 - by_1)}{a^2 + b^2}.$$

3. If Q and Q' are two distinct points with the same respective distances from three points P_1, P_2, and P_3, then P_1, P_2, and P_3 lie on the equidistant line of Q and Q'. Hence if P_1, P_2, and P_3 are not in a line there can be only one point Q with given distances from them. □

As an example of the first property, if $P = (-a, 0)$ and $Q = (a, 0)$ then a point (x, y) is equidistant from P and Q if and only if

$$(x + a)^2 + y^2 = (x - a)^2 + y^2,$$

whence

$$x = 0,$$

which is the equation of the axis OY.

Another advantage of this definition of the Euclidean plane is that it admits a concept of "moving" one figure until it coincides with another, as in Pappus' proof that the base angles of an isosceles triangle are equal (Section 2.4). To formalize this idea, we consider functions that "preserve distance."

Definition A function f on $\mathbb{R} \times \mathbb{R}$ is an *isometry* (from the Greek for "same distance") if $d(f(P_1), f(P_2)) = d(P_1, P_2)$ for any two points P_1 and P_2.

An example of an isometry is the function ref_{OY} that sends each point (x, y) to $(-x, y)$. We call this function *reflection in OY* (hence the symbol ref_{OY}) because it captures the intuitive idea of mirror reflection in the line OY. It preserves distances, because if $P_1 = (x_1, y_1)$ and $P_2 = (x_2, y_2)$ then $\text{ref}_{OY}(x_1, y_1) = (-x_1, y_1)$ and $\text{ref}_{OY}(x_2, y_2) = (-x_2, y_2)$, hence

$$d(\text{ref}_{OY}(P_1), \text{ref}_{OY}(P_2)) = \sqrt{(-x_2 + x_1)^2 + (y_2 - y_1)^2}$$
$$= \sqrt{(x_2 - x_1)^2 + (y_2 - y_1)^2}$$
$$= d(P_1, P_2).$$

Now suppose we have a triangle ABC with $CA = CB$. We can re-create Pappus' proof by placing the triangle with A and C on OX, with O at their midpoint, say $A = (-a, 0)$, and $C = (a, 0)$. Because C is equidistant from A and B, it must be on OY by the preceding calculation. If we then reflect triangle ABC in OY, it is mapped onto itself. In particular, the angle at A is mapped onto the angle at C; hence these two angles are equal.

Proving that two angles are equal can usually be done, as here, by moving one to coincide with the other. Actually measuring angles is harder, but it can also be done with the help of the real numbers, as we shall see in Chapter 5.

Exercises

Another useful isometry is the *half turn*, or rotation through π. The half turn about O is the function $\text{rot}_{O,\pi}$ that sends (x, y) to $(-x, -y)$.

3.5.1. Check that the half turn about O is an isometry, and use it to show that vertically opposite angles between lines through O are equal.

We can prove that vertically opposite angles are equal at any point (a, b) with the help of an isometry that moves O to (a, b). The simplest such isometry is the *translation* $\text{tran}_{a,b}$, which moves each point (x, y) to the point $(x + a, y + b)$. It is reversed by the translation $\text{tran}_{-a,-b}$, which sends (x, y) to $(x - a, y - b)$.

3.5.2. Check that $\text{tran}_{a,b}$ is an isometry.

3.5.3. Show that vertically opposite angles at any point (a, b) are equal by translating (a, b) to O, applying a half turn, then translating the angles back to (a, b).

Other classical results about equal angles can also be proved by using isometries to formalize intuitive movements of one angle onto another. An example is the pair of *alternate* angles that occur where a line crosses two parallels (Figure 3.7).

3.5.4. Prove that alternate angles are equal by a suitable combination of translations and half turns.

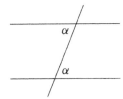

FIGURE 3.7   Alternate angles.

3.5.5. Deduce from Exercise 3.5.4 a proof that the angle sum of any triangle is two right angles.

3.6   Isometries of the Euclidean Plane

Isometries of the Euclidean plane are actually not much more complicated than the example in the previous section, though this is not clear from the bare definition. The situation is crucially simplified by the following.

Basic property of isometries   *An isometry is determined by the images of three points not in a line.*

Proof   This theorem is based on the third basic property of distance, that each point is determined by its distances from three points not in a line, but first we have to show that the image of any isometry includes three such points.

Three particular points not in a line are the vertices A, B, C of an equilateral triangle. C is not in the line AB because AB is the equidistant set of two points C and D (Figure 3.8), and C is certainly not equidistant from C and D. Notice that this argument depends only on the distances between A, B, C, and D, and these distances

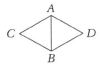

FIGURE 3.8   Points not in a line.

are preserved by any isometry. Hence if f is any isometry, and A, B, C are the vertices of an equilateral triangle, then $f(A)$, $f(B)$, $f(C)$ are three points not in a line.

If Q is any point, we know that Q is determined by its distances from A, B, and C. Its image $f(Q)$ has the same distances from $f(A)$, $f(B)$, and $f(C)$, respectively, and because $f(A)$, $f(B)$ and $f(C)$ are not in a line, there is only one point with the same distances from them as $f(Q)$. Thus any isometry that agrees with f on A, B, and C agrees with f on any point Q.   □

This theorem will allow us to express any isometry f of the Euclidean plane as a composite of simple isometries, if only we can find enough isometries to move A, B, C to $f(A)$, $f(B)$, $f(C)$, respectively. The most convenient isometries for this purpose arise from the fact that each line is the equidistant set of two points. The calculations in the previous section show that the line $ax + by = c$ is the equidistant set of any two points (x_1, y_1), (x_2, y_2) that satisfy the relation

$$x_2 = x_1 + \frac{k}{2}a, \qquad y_2 = y_1 + \frac{k}{2}b,$$
$$\text{where} \qquad k = \frac{4(c - ax_1 - by_1)}{a^2 + b^2}.$$

Because these points can be regarded as "mirror images" in the line $ax + by = c$, it is reasonable to make the following definition.

Definition   *Reflection in the line* $ax + by = c$ *is the map that sends each point* (x_1, y_1) *to the point* (x_2, y_2) *defined by*

$$x_2 = x_1 + \frac{2a(c - ax_1 - by_1)}{a^2 + b^2},$$
$$y_2 = y_1 + \frac{2b(c - ax_1 - by_1)}{a^2 + b^2}.$$

It follows that any two points can be exchanged by reflection in their equidistant line, but we have to check that reflection in $ax + by = c$ is an isometry. This is easier if we first arrange that $a^2 + b^2 = 1$, which can always be done because an equivalent equation is obtained if a, b, and c are multiplied by any nonzero constant. The reflection that sends (x_1, y_1) to (x_2, y_2) is then expressed by the equations

$$x_2 = x_1 + 2a(c - ax_1 - by_1) = x_1(1 - 2a^2) - 2aby_1 + 2ac,$$
$$y_2 = y_1 + 2b(c - ax_1 - by_1) = y_1(1 - 2b^2) - 2abx_1 + 2bc.$$

Let $P_1 = (x_1, y_1)$ and $P_1' = (x_1', y_1')$ be any two points, and consider the square of the distance between their reflections P_2 and P_2' in $ax + by = c$:

$$\left[(x_1' - x_1)(1 - 2a^2) - (y_1' - y_1)2ab\right]^2$$
$$+ \left[(y_1' - y_1)(1 - 2b^2) - (x_1' - x_1)2ab\right]^2.$$

Expanding the two main terms, one finds that the coefficients of $(x_1' - x_1)^2$, $(x_1' - x_1)(y_1' - y_1)$, and $(y_1' - y_1)^2$ are 1, 0, and 1, respectively, because $a^2 + b^2 = 1$. Hence $d(P_2, P_2') = d(P_1, P_1')$ as required. $\quad\square$

The work involved in formalizing the intuitively simple idea of reflection in a line is worthwhile, because it gives us *all* isometries of the Euclidean plane. We get them as *composites* of reflections, that is, as the result of successive reflections.

Three reflections theorem   *Each isometry of the Euclidean plane is the composite of one, two, or three reflections.*

Proof   Suppose that P_1, P_2, and P_3 are three points not in a line, and f is any isometry. By the basic property of isometries, it suffices to find a composite of one, two, or three reflections that send P_1, P_2, and P_3 to $f(P_1)$, $f(P_2)$, and $f(P_3)$ respectively, because the latter isometry necessarily coincides with f. This can be done with the following reflections f_1, f_2, and f_3.

1. Let f_1 be reflection in the equidistant line of P_1 and $f(P_1)$. It sends P_1 to $f(P_1)$ (by definition of reflection), P_2 to $f_1(P_2)$, and P_3 to $f_1(P_3)$.

2. If $f_1(P_2) \neq f(P_2)$, let f_2 be reflection in the equidistant line of $f_1(P_2)$ and $f(P_2)$. Then f_2 sends $f_1(P_2)$ to $f(P_2)$, as required. Also $f_1(P_1) = f(P_1)$ is equidistant from $f_1(P_2)$ and $f(P_2)$ (namely, at the distance between P_1 and P_2), hence it is fixed by the reflection f_2.

3. We now have $f_2f_1(P_1) = f(P_1)$ and $f_2f_1(P_2) = f(P_2)$. If $f_2f_1(P_3) \neq f(P_3)$ we send $f_2f_1(P_3)$ to $f(P_3)$ by reflection f_3 in their equidistant line. Again, $f(P_1) = f_2f_1(P_1)$ is equidistant from $f(P_3)$ and $f_2f_1(P_3)$, hence it is fixed by f_3. So is $f(P_2) = f_2f_1(P_2)$.

It follows that P_1, P_2, and P_3 are sent to $f(P_1)$, $f(P_2)$, and $f(P_3)$, respectively, by either f_1, f_2f_1, or $f_3f_2f_1$. Thus one of these composites of reflections is the required isometry. $\qquad\square$

While this proof is in front of you, two aspects of terminology and notation should be pointed out.

- We speak of the "composite" of reflections because reflections are functions, and taking a "function of a function" is called *composition*. The notation for composition of functions (for example, $f_2f_1(P_1)$) is the usual product notation, but we prefer not to call this a "product" because it does not have all the properties of other products.

- In particular, the composite f_2f_1 is not necessarily the same as f_1f_2. For example, let f_1 be reflection in the x-axis and let f_2 be reflection in the line $x = y$; then f_2f_1 is a quarter turn anticlockwise and f_1f_2 is a quarter turn clockwise. So don't forget that f_2f_1 means "f_1 first, then f_2."

Exercises

Several general properties of isometries follow from the three reflections theorem, because they are easily proved for reflections.

3.6.1. Show that the following are true of reflections, and hence of all isometries.

1. They are invertible functions, and their inverses are also isometries.

2. They map lines to lines.

The three reflections theorem should also account for the isometries we already know from Section 3.5, the half turn about O and the translations.

3.6.2. Show that a half turn about O is the composite of reflections in OX and OY.

3.6.3. Show that the translation $\mathrm{tran}_{a,b}$ is the composite of reflection in the equidistant line of O and (a, b) and reflection in the parallel to this line through (a, b).

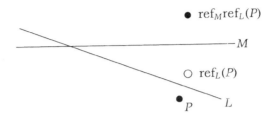

FIGURE 3.9   Rotation via reflections in intersecting lines.

The half turn is an example of a *rotation*, which is defined in general to be the composite of reflections in intersecting lines. Because reflection in a line L leaves all points of L fixed (this is clear from the defining formulas), the composite of reflections in lines L and M leaves the intersection of L and M fixed. A picture also suggests that this rotation moves any other point through *twice* the angle between L and M (Figure 3.9).

3.6.4. Show that the composite $\mathrm{ref}_M\mathrm{ref}_L$ of reflections ref_L in L and ref_M in M moves the line L through twice the angle between L and M.

3.6.5. Show that the rotation about O obtained by successive reflections in OX and the line $y = mx$ sends (x, y) to $(\frac{1-m^2}{1+m^2}x - \frac{2m}{1+m^2}y,\ \frac{2m}{1+m^2}x + \frac{1-m^2}{1+m^2}y)$.

You may recognize this as the standard formula for "rotation through angle θ," where $m = \tan\frac{\theta}{2}$, $(1 - m^2)/(1 + m^2) = \cos\theta$, and $2m/(1 + m^2) = \sin\theta$. The same formulas will recur when we study rational points on the circle in the next chapter.

There is one more type of Euclidean isometry that is not a reflection, translation, or rotation. It is called a *glide reflection*, and it is the composite of a translation with a reflection in a line parallel to the direction of translation.

3.6.6.* Show that any composite of three reflections is a glide reflection.

3.7   The Triangle Inequality

Another crucial property of distance is the so-called triangle inequality: *if A, B, and C are three points not in a line then*

$$d(A, C) < d(A, B) + d(B, C).$$

This property is so obvious that even a dog knows it; a dog will go straight from A to a bone at C rather than go via B. However, in mathematics, even obvious statements need not be accepted if they can be proven. It so happens that the triangle inequality *can* be proved from the standard assumptions of geometry, though perhaps not as easily as one would like.

Euclid arrives at the triangle inequality only in his Proposition 20 (*Elements*, Book I), and it depends on most of his earlier propositions. A proof in our setup is also not obvious, but it takes only a few lines of algebra. We can simplify the calculation by applying isometries to move triangle ABC to a convenient position. First we apply a translation to move B to the origin. Then, if A is not already on the x-axis, we exchange it with the point A' on the x-axis such that $d(A', B) = d(A, B)$, by reflecting in their equidistant line.

The result is a triangle with coordinates of the form

$$A = (a_1, 0), \quad B = (0, 0), \quad C = (c_1, c_2),$$

with $c_2 > 0$ if the three points are not in a line. The required inequality

$$d(A, C) < d(A, B) + d(B, C)$$

then takes the form

$$\sqrt{(c_1 - a_1)^2 + c_2^2} < a_1 + \sqrt{c_1^2 + c_2^2},$$

and it is true because

$$(\text{RHS})^2 - (\text{LHS})^2 = a_1^2 + 2a_1\sqrt{c_1^2 + c_2^2} + c_1^2 + c_2^2 - (c_1 - a_1)^2 - c_2^2$$

$$= 2a_1\sqrt{c_1^2 + c_2^2} - 2a_1 c_1$$

$$> 0,$$

because $c_2 > 0$ and therefore $\sqrt{c_1^2 + c_2^2} > c_1$.   □

This calculation also shows that $d(A, C) = d(A, B) + d(B, C)$ only when $c_2 = 0$, that is, when the three points are a line.

One reason Euclid's proof of the triangle inequality is longer than ours is that he assumes less. He proves it without assuming the parallel axiom, so his argument also applies to the geometry of the non-Euclidean plane (Section 3.9*).

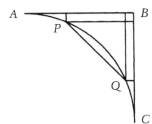

FIGURE 3.10 Bounding the length of polygons in the circle.

Exercises

A less formal way to express the triangle inequality is by the old saying "a straight line is the shortest distance between two points."

3.7.1. Use the triangle inequality and induction to prove that a line segment AB is the shortest *polygonal path* from A to B. (A polygonal path is a sequence of line segments $A_1A_2, A_2A_3, \ldots, A_{n-1}A_n$.)

It follows from this that the line segment is also the shortest *curve* from A to B, because we define the length of a curve K to be the least upper bound of lengths of polygonal paths from A to B with their vertices on K, provided the least upper bound exists.

For certain curves, such as the circle, we can also prove the existence of an upper bound by the triangle inequality. It then follows that the least upper bound exists, by the completeness of the real numbers (Exercises 3.4.1 and 3.4.2).

3.7.2. Deduce from Figure 3.10 and the triangle inequality that any polygonal path with vertices on the quarter circle from A to C has total length $< d(A, B) + d(B, C)$.

3.8* Klein's Definition of Geometry

The history of geometry is a story with a shifting point of view. In Euclid's time the raw materials of geometry were points, lines, and planes, and theorems were proved from visual axioms with the help of constructions and the vague idea of "movement." Numbers had a very limited role, because irrational lengths were not believed to correspond to numbers. In the 17th century, Fermat and Descartes introduced numbers as coordinates and used algebra to simplify the description and manipulation of figures. However, coordinates were just a means to describe figures; they were not considered geometric in themselves. It was not until Dedekind and others clarified the concept of number in the late 19th century that numbers could be seen as the raw material for creating points, lines, and planes. This was virtually a reversal of Euclid's viewpoint, and it included a reversal of the role of the Pythagorean theorem—from a theorem about triangles to the definition of distance between two points in the plane.

Other basic theorems of Euclid's geometry can be proved, in this new setup, with the help of the concept of isometry, which is a rigorous counterpart of Euclid's idea of "movement."

In 1872, Klein made yet another dramatic shift in viewpoint when he realized that *the isometries make the geometry*. In particular, Euclidean plane geometry is *everything that is preserved by Euclidean isometries*. Indeed, the fundamental quantity preserved by Euclidean isometries is the distance $d(P_1, P_1')$ between points. We saw in Section 3.6 that any reflection preserves distance and that any isometry is a composite of reflections, hence $d(P_1, P_1')$ is preserved by all isometries. In principle, we could start with the set of reflections and "discover" the idea of Euclidean distance $d(P_1, P_1')$, by calculating the quantity

$$\sqrt{(x_2' - x_2)^2 + (y_2' - y_2)^2},$$

for the mirror images $P_2 = (x_2, y_2)$ and $P_2' = (x_2', y_2')$ of points $P_1 = (x_1, y_1)$ and $P_1' = (x_1', y_1')$ under any reflection, and finding it equal to the corresponding quantity for $P_1 = (x_1, y_1)$ and $P_1' = (x_1', y_1')$,

$$\sqrt{(x_1' - x_1)^2 + (y_1' - y_1)^2}.$$

As we know, the concepts of *line* (the equidistant set of two points) and *circle* (the equidistant set of one point) can be defined in terms of distance, so as soon as the distance function is derived from the isometries, we have the whole Euclidean plane geometry.

Thus Klein's idea gives yet another way to put Euclid's geometry on a rigorous foundation. As usual, the test of a new viewpoint is whether it enables us to see anything more clearly than before.

Klein's viewpoint shows us that Euclidean geometry is just one of several structurally similar geometries. One of them, of course, is the geometry of Euclidean *space,* for which the isometries are composites of reflections in planes.

Among the geometries of surfaces, the most familiar relative of Euclidean geometry is the geometry of the sphere. Its isometries are composites of reflections in planes through the sphere's center. The "equidistant sets" of these reflections are the intersections of the planes with the sphere, the so-called *great circles.* These are the "lines" of spherical geometry, and their basic properties are found by arguments similar to those we used for the Euclidean plane.

1. Any two "lines" have a point in common.

2. Hence the composite of two reflections is always a rotation.

3. Any isometry is the composite of one, two or three reflections.

An important part of Klein's concept of geometry is that the isometries form a *group of transformations,* a set of one-to-one functions closed under composites and inverses. Such a set is obtained by taking composites of reflections because each reflection is its own inverse; that is, the composite of a reflection with itself is the *identity function,* which sends each point to itself. It follows that each composite $f_1 f_2 \cdots f_k$ of reflections also has an inverse, namely, the composite (in reverse order) of their inverses, $f_k^{-1} \cdots f_2^{-1} f_1^{-1}$. Because if we compose these two composites we get

$$f_1 f_2 \cdots f_{k-1} f_k \cdot f_k^{-1} f_{k-1}^{-1} \cdots f_2^{-1} f_1^{-1} = f_1 f_2 \cdots f_{k-1} \cdot f_{k-1}^{-1} \cdots f_2^{-1} f_1^{-1}$$

$$\vdots$$

$$= f_1 f_2 \cdot f_2^{-1} f_1^{-1}$$
$$= f_1 \cdot f_1^{-1}$$
$$= \text{identity function}$$

by successive cancellation of inverses. Thus isometries of both the Euclidean plane and the sphere have inverses, and hence the corresponding sets of isometries are groups.

Viewing the set of isometries as a group draws our attention to *subgroups*—subsets of isometries that also form groups—and these also throw new light on geometry.

For example, consider the isometries that are composites of an *even* number of reflections. The inverse of such an isometry is also the composite of an even number of reflections, and so is the composite of two such isometries. Thus the composites of even numbers of reflections form a subgroup. Intuitively speaking, it is the subgroup that preserves "orientation," or "handedness," or "clock-wiseness" or whatever you want to call those aspects of geometry that are *not* preserved by reflections. One such aspect is the cyclic order of the numbers 1, 3, 6, 12 on a clock face, or at least it seems to be. We can escape the tricky problem of defining these aspects by *letting the subgroup define them.* That is, we say that a property *depends on orientation* if it is not preserved by the whole group but preserved by the subgroup of composites of even numbers of reflections. We call the latter group the *orientation-preserving subgroup.*

This idea depends on the fact that the orientation-preserving subgroup is not the whole isometry group; it would fail if a composite of an even number of isometries was also a composite of an odd number of reflections. The example of the clock face makes this unlikely, but we can prove it by considering composites of one or two reflections and what they do to lines. A single reflection maps one line onto itself. A composite of two reflections is either a rotation, which maps no line onto itself, or else a translation, which maps infinitely many lines onto themselves. Thus a reflection cannot be a composite of two reflections, and it follows that it cannot be a composite of any even number of reflections, because all such products are rotations or translations (see exercises). Thus the orientation-preserving subgroup is not the whole isometry group, and hence there is such a thing as orientation!

Exercises

One reason that composites of even numbers of reflections are rotations or translations is that a rotation is the composite of reflections in two intersecting lines. It appears intuitively clear (see Figure 3.9 again) that the rotation depends only on the point of intersection and the angle between the lines, hence it should be possible to represent any rotation

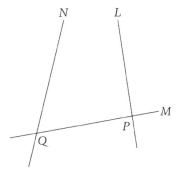

FIGURE 3.11   Reflection lines for the composite of two rotations.

about P as the composite of reflections in two lines, *one of which is given.* The first exercise confirms this intuition.

3.8.1. Let L, M, and L' be any lines through a point P. Show, using the three reflections theorem or otherwise, that there is a line M' through P such that

$$\text{ref}_M \text{ref}_L = \text{ref}_{M'} \text{ref}_{L'}.$$

Also prove that if M' is given, we can find L' to satisfy this equation.

3.8.2. Deduce from Exercise 3.8.1 that rotations about two points P and Q can be expressed as $\text{ref}_M \text{ref}_L$ and $\text{ref}_N \text{ref}_M$, where L, M, and N meet the points P and Q as indicated in Figure 3.11.

So far, these arguments apply to both the Euclidean plane and the sphere. The possibility of a translation arises in the next exercise, but only in the Euclidean plane.

3.8.3. Conclude from Exercise 3.8.2 that the composite of rotations is either a rotation or a translation. If it is a rotation, about which point?

It follows that, on the sphere, the composite of two rotations is a rotation, and hence the orientation-preserving subgroup of the sphere consists entirely of rotations. In the Euclidean plane, we still have to find how translations interact with each other and with rotations.

3.8.5. Show that the composite of translations is a translation.

3.8.6. By imitating the arguments in Exercises 3.8.1 and 3.8.3, or otherwise, show that the composite of a translation and a rotation is either a translation or a rotation.

3.9* The Non-Euclidean Plane

A beautiful example of the way isometries create geometry is the non-Euclidean plane of Henri Poincaré (1882). Poincaré found a geometry in which there is more than one parallel to a given "line" through a given point. His "plane" is the upper half ($y > 0$) of $\mathbb{R} \times \mathbb{R}$, and his "reflection" is a generalization of ordinary reflection called *reflection in a circle*. The non-Euclidean *isometries* are composites of reflections, and non-Euclidean *distances* are equal if there is a non-Euclidean isometry carrying one to the other.

The reflection of a point P in a circle C with center Z and radius r is defined to be the point P' on the Euclidean line ZP such that

$$ZP \cdot ZP' = r^2.$$

See Figure 3.12. Ordinary reflection can be regarded as the limiting case of reflection in a circle as the center Z tends to infinity.

In fact, the reflections generating the isometries of the non-Euclidean plane include ordinary reflections in the vertical lines $x = $ constant. The other reflections used are reflections in circles with their centers on the x-axis. Thus the "lines" of the non-Euclidean plane are obtained immediately as the fixed point sets of the "reflections," namely, the vertical Euclidean half-lines $x = $ constant and the Euclidean semicircles with centers on the x-axis (Figure 3.13).

We can see from Figure 3.13 that the parallel axiom fails: $\mathcal{M}$ and $\mathcal{N}$ are two "lines" through the point P that do not meet the "line" $\mathcal{L}$.

Apart from the parallel axiom, all other axioms of Euclid's geometry hold in the non-Euclidean plane. For example, there is exactly one "line" through any two points P and Q, and in fact it can be found by a ruler and compass construction, as Figure 3.14 shows. One draws the Euclidean line PQ and, if PQ is not vertical, con-

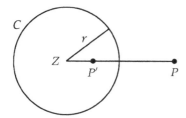

FIGURE 3.12 Reflection of a point in a circle.

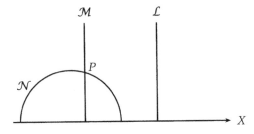

FIGURE 3.13 Some "lines" of the non-Euclidean plane.

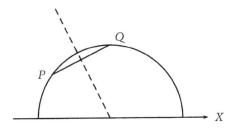

FIGURE 3.14 Construction of a non-Euclidean "line."

structs its perpendicular bisector. The latter meets the x-axis at the center of the semicircle that is the non-Euclidean "line" PQ.

Another pleasant property of Poincaré's non-Euclidean plane is that its "angles" are ordinary angles. The only difference is that they are not angles between Euclidean lines but between "lines," that is, between Euclidean circles, or between a circle and a vertical Euclidean line. The angle between circles is the angle between their tangents at the point of intersection. For example, perpendicular "lines" are either perpendicular semicircles with their centers on the x-axis, or a vertical Euclidean half-line and a semicircle with its center at the lower end of the half-line (see Figures 3.17 and 3.16).

It turns out that a non-Euclidean "circle"—the set of points at constant non-Euclidean distance from a single point—is a Euclidean circle. Its Euclidean center is *not* the same as its non-Euclidean center (because Euclidean distance is not the same as non-Euclidean distance, as we shall see). However, the circle can also be constructed from its non-Euclidean center and a point on its circumference by ruler and compass. This means that the natural non-Euclidean "constructions" can all be done within Euclidean geometry. In particular, non-Euclidean "constructible points" are constructible points of the

Euclidean half-plane. Some of these constructions are pursued in the exercises.

The isometries of the non-Euclidean plane are particularly interesting. Those that preserve orientation are composites of two reflections, and there are three types, as the "lines" of reflection meet at a point of the half-plane, meet on the x-axis, or do not meet at all. The first are non-Euclidean *rotations,* the second are called *limit rotations* (because the center of rotation is infinitely far away, in terms of non-Euclidean distance), and the third are non-Euclidean *translations.*

An example of a non-Euclidean "translation" is the composite of reflections in the semicircles with center O and radii 1 and 2. It is easy to check that this "translation" sends each (x, y) in the upper half-plane to $(4x, 4y)$. Hence each of the points $x = 0$, $y = 1, 1/4, 1/4^2, 1/4^3, \ldots$ is mapped onto its predecessor by this translation, and so they are *equally spaced* in the sense of non-Euclidean distance. This explains why the x-axis is infinitely far away from all points of the non-Euclidean plane and shows that non-Euclidean lines are infinitely long.

This may seem to be a strange geometry, with semicircles called "lines" of infinite "length," but because all but one of Euclid's axioms hold it is feasible to use ordinary geometric reasoning. Poincaré, in fact, introduced this non-Euclidean plane because he wanted to study transformations generated by reflections in circles, and he found that geometric language made them easier to understand.

Exercises

Reflection in a circle occurs frequently in mathematics, and it can be described in many ways. Its connection with ruler and compass constructions is established by Figure 3.15.

3.9.1. By comparison of similar right-angled triangles, show that $ZP \cdot ZP' = r^2$, and hence describe a ruler and compass construction of P' from P, and P from P'.

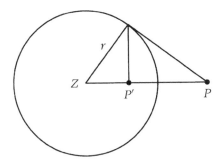

FIGURE 3.15 Construction of reflection in a circle.

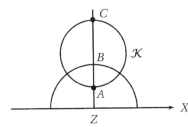

FIGURE 3.16 Perpendicular diameters of a non-Euclidean circle.

The most important properties of reflection in a circle are that it preserves circles and angles, hence the angles and circles in the non-Euclidean plane look the same as Euclidean angles and circles. (Except that non-Euclidean angles usually occur between circles rather than between Euclidean straight lines.) We shall assume these facts in the exercises that follow. Their aim is to construct any non-Euclidean circle, given its non-Euclidean center B and a point A on its circumference. The simplest case is shown in Figure 3.16, where A and B are on a vertical Euclidean line.

3.9.2. Show that the reflection C of A in the semicircle through B with center at Z is the point opposite A on the non-Euclidean circle $\mathcal{K}$ with radius BA. Hence give a ruler and compass construction of $\mathcal{K}$.

The general case is where A and B are not on the same vertical line (Figure 3.17). In this case we first construct the "line" AB (semicircle through A and B).

3.9.3. Describe a ruler and compass construction of the "line" $\mathcal{L}$ through B perpendicular to the "line" AB.

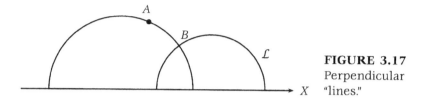

FIGURE 3.17
Perpendicular
"lines."

Now we construct the reflection C of A in $\mathcal{L}$. This is the point opposite A on the non-Euclidean circle $\mathcal{K}$ with radius BA. $\mathcal{K}$ itself is therefore a Euclidean circle through A and C perpendicular to the semicircle through A and C (Figure 3.18).

3.9.4. Describe a ruler and compass construction of the Euclidean center of this circle.

This completes the proof that non-Euclidean "ruler and compass" constructions can be done by Euclidean ruler and compass. It follows that if we take, say, the line segment from $(0, 1)$ to $(0, 2)$ as the non-Euclidean unit of length, then all non-Euclidean constructible points have coordinates expressible by rational operations and square roots (by Section 3.2). In particular, it is impossible to construct the point $(0, \sqrt[3]{2})$, by Exercises 3.2.5* to 3.2.8*. This leads to a surprising conclusion.

3.9.5. The line segment from $(0, 1)$ to $(0, \sqrt[3]{2})$ has $1/3$ the non-Euclidean length of the line segment from $(0, 1)$ to $(0, 2)$. Why? Deduce that trisection of a "line" segment by "ruler and compass" is not always possible in the non-Euclidean plane.

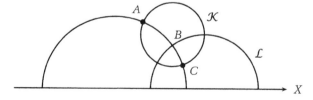

FIGURE 3.18
Non-Euclidean
circle and
diameters.

3.10 Discussion

Algebra and Geometry

The main message of the last two chapters is that geometry has two sides. First there is the visual, self-contained, synthetic side, which seems intuitively natural; then the algebraic, analytic side, which takes over when intuition fails and integrates geometry into the larger world of mathematics. Visualization will surely continue to inspire new discoveries in geometry, but it is equally likely that algebra will rule geometry as long as it is more efficient and more conducive to mathematical unity. A rigorous synthetic development of geometry requires too many complicated axioms that, unlike those of algebra, are not easily used in other parts of mathematics.

A second important message is that there is an apparent conflict between these two sides, and that this conflict has been very fruitful for the development of mathematics. As we know, the conflict began with the discovery of irrational lengths, such as $\sqrt{2}$. At the time, numbers were rational by definition, so irrational lengths could *not* be numbers, and hence geometry could not be based on arithmetic. Doing geometry without arithmetic turned out to be fruitful, however, because it led to Euclid's *Elements*, the most influential mathematics book of all time. In fact, it was only when Euclid's influence began to wane, in the 19th century, that mathematicians finally considered resolving the conflict between arithmetic and geometry, by extending the concept of number.

The latter development was also extraordinarily fruitful, as completion of the number concept not only clarifies the nature of points and lines, but also of curves and other objects too complicated to be grasped by Euclid's methods. For example, we shall see in Chapters 5 and 9 how completeness of the real numbers enables us to define lengths of curves and areas of curved regions.

The 19th century also saw a great enlargement in the scope of algebra, which allowed the operations $+$, $-$, $\times$, and $\div$ to be applied to objects that are not necessarily numbers. Of course, it is helpful to use the symbol $+$ only for a function that behaves like ordinary addition on numbers, so first it was necessary to find the characteristic properties of ordinary $+$, $-$, $\times$, and $\div$, and to describe them as simply as possible. This led to the definitions of *ring* and *field*, whose characteristic properties were listed in Section 1.4.

At this stage (around 1900) it became clear that the concept of *field* was the appropriate algebraic setting for geometry. We have already seen (in Sections 3.1 to 3.5) how to build Euclid's geometry using the field $\mathbb{R}$ of real numbers. Conversely, it is possible to build a field from Euclid's geometric concepts. The field consists of "positive" elements l and their additive inverses $-l$, and each l is a "length." Lengths are added, multiplied, and divided using the constructions of Exercises 3.2.1 and 3.2.2, and it follows from the axioms of geometry that the lengths and their additive inverses indeed form a field. In fact, if we also assume completeness, the field turns out to be nothing but $\mathbb{R}$.

Thus doing geometry analytically, using real number coordinates, is *almost* equivalent to doing geometry synthetically. The only extra ingredient in analytic geometry is completeness, which amounts to assigning a number to each point on the line. This of course is precisely the step the Greeks refused to take. Does this mean they missed analytic geometry only by a whisker, because of their scruples over irrationals, or was the concept of $\mathbb{R}$ really remote from Greek mathematics?

The Jump from $\mathbb{Q}$ to $\mathbb{R}$

Dedekind's definition of real numbers as cuts in $\mathbb{Q}$ is exquisitely simple, but deceptive in a way, because it hides the fact that $\mathbb{R}$ is a far less comprehensible set than $\mathbb{Q}$. The set $\mathbb{Q}$ is simpler than it looks, being similar in character to the set $\mathbb{N}$. Of course, $\mathbb{N}$ is an infinite set, but we can comprehend it as the result of the process of starting with 1 and repeatedly adding 1. It is not necessary to imagine this process actually *completed*, only continued indefinitely, to grasp the meaning of $\mathbb{N}$, because any *particular* member of $\mathbb{N}$ is reached after a finite amount of time. The infinite set $\mathbb{Z}$ can be grasped in the same way, as the result of a process that starts with 0 and alternately adds 1 and changes sign. This is why it makes sense to write

$$\mathbb{Z} = \{0, 1, -1, 2, -2, 3, -3, \dots\}.$$

The process that generates the list $0, 1, -1, 2, -2, 3, -3, \ldots$ is clear, and it is also clear that any member of $\mathbb{Z}$ will eventually appear.

To capture the rationals by a list-generating process is only slightly more complicated. First we generate the positive rationals, by the process that produces them in the following order:

$\frac{1}{1}$, (reduced fractions whose top and bottom line sum to 2)

$\frac{2}{1}, \frac{1}{2}$, (reduced fractions whose top and bottom line sum to 3)

$\frac{3}{1}, \frac{1}{3}$, (reduced fractions whose top and bottom line sum to 4)

$\frac{4}{1}, \frac{3}{2}, \frac{2}{3}, \frac{1}{4}$, (reduced fractions whose top and bottom line sum to 5)

$\frac{5}{1}, \frac{1}{5}$, (reduced fractions whose top and bottom line sum to 6)

$\vdots$

Then $\mathbb{Q}$ itself can be listed by alternating positive and negative numbers the way we did with $\mathbb{Z}$. Thus, when viewed merely as an infinite set, $\mathbb{Q}$ is just as comprehensible as $\mathbb{N}$.

The sets $\mathbb{N}$, $\mathbb{Z}$, and $\mathbb{Q}$ are called *countable* because the listing processes give each of them a first member, second member, third member, and so on.   And because the members of $\mathbb{Z}$ or $\mathbb{Q}$ are thereby paired with members of $\mathbb{N}$ (first member with 1, second member with 2 and so on), all three sets are reckoned to be the "same size" or, as we say, the same *cardinality*.

It was a great surprise to mathematicians when Cantor discovered in 1874 that *not* all infinite sets have the same cardinality. In particular, $\mathbb{R}$ is *uncountable*, and hence of greater cardinality than $\mathbb{N}$, because there is no way to list its members. This means that $\mathbb{R}$ can only be comprehended (if at all) as a *completed whole*. To understand $\mathbb{R}$, we not only have to grasp the individual cuts in $\mathbb{Q}$, we have to grasp them all at once!

This is what makes the concept of $\mathbb{R}$ really remote from Greek mathematics. The Greeks were willing to accept a "potential" infinity, such as the process for generating $\mathbb{N}$, but not the "actual" infinity of the set $\mathbb{N}$ itself. This was no obstacle to elementary number theory, because they could speak of an "arbitrary natural number" instead of the set $\mathbb{N}$. But they could not speak of an arbitrary *real* number, because this presupposes infinite subsets L and U of $\mathbb{Q}$, given without generating processes, and hence actually infinite.

Interestingly, the reason $\mathbb{R}$ can be comprehended only as a completed whole is precisely its completeness in the mathematical sense. We can, in fact, prove that any countable set A of numbers contains a gap, and hence is not complete. The idea is to use a list of members of A to find sequences $b_1, b_2, b_3, \ldots$ and $c_1, c_2, c_3, \ldots$ of members with

$$b_1 < b_2 < b_3 < \cdots < c_3 < c_2 < c_1,$$

and no member of A between the b_is and c_is. Then the separation of A into the set B of numbers $\leq$ some b_i and the set C of numbers $\geq$ some c_j gives a gap. Either the sequences of b_is and c_is are finite and there is gap between the greatest b_i and the least c_j; or else there is no greatest b_i and no least c_j, so (B, C) itself is a gap in $\mathbb{R}$. Given the job of separating A, the sequences $b_1, b_2, b_3, \ldots$ and $c_1, c_2, c_3, \ldots$ virtually define themselves.

Let $a_1, a_2, a_3, \ldots$ be a list of members of A, and let

$$b_1 = \text{first number on the list that is not the}$$
$$\text{maximum member of } A;$$
$$c_1 = \text{first number on the list that is } > b_1.$$

Notice that this means b_1 is one of a_1 or a_2, and c_1 is the other. The remaining numbers $b_2, c_2, b_3, c_3, \ldots$ are chosen in that order by looking at $a_3, a_4, a_5, \ldots$ in turn and letting

$$b_{n+1} = \text{first } a_k > b_1, b_2, \ldots, b_n \text{ and } < c_1, c_2, \ldots, c_n;$$
$$c_{n+1} = \text{first } a_k > b_1, b_2, \ldots, b_{n+1} \text{ and } < c_1, c_2, \ldots, c_n.$$

It follows immediately that there is no a_k between the sequences $b_1, b_2, b_3, \ldots$ and $c_1, c_2, c_3, \ldots$. If there were, we would look at it at some stage and *make it a member* of one of these sequences, which is a contradiction.

This argument is essentially the one given by Cantor himself in 1874. He later gave a more popular argument, called the *diagonal argument*, which is based on decimal expansions of the real numbers. The current argument is more elementary, however, and better suited to the point we wish to make—that the completeness of $\mathbb{R}$ implies its uncountability.

Incidentally, Cantor's argument gives another way to show that irrational numbers exist. If we take A to be the set of rationals, then (B, C) is a cut defining an irrational.

A Different Definition of Euclidean Geometry

One of the characteristic features of Euclidean plane geometry is the existence of *similarities*: mappings of $\mathbb{R} \times \mathbb{R}$ that multiply all lengths by a constant. A typical similarity is the *dilatation* dil_c (for $c \neq 0$) that sends each point (x, y) to the point (cx, cy) and consequently multiplies all lengths by c. We take the existence of similarities for granted in real life, in assuming that scale models, maps, and so on are faithful representations of real objects.

At the same time, the existence of scale models means that length is not really an essential concept in Euclidean geometry. The important properties of a triangle, for example, are not the lengths of its sides, but the *ratios* of the lengths (which determine the angles of the triangle). In fact, the theorems of Euclid's geometry are really about ratios of lengths, not about lengths themselves.

For this reason, we might very well define Euclidean geometry by using the group of similarities of $\mathbb{R} \times \mathbb{R}$ rather than the group of isometries. Similarities do indeed form a group, because each similarity is the composite of a dilatation dil_c with an isometry f, and its inverse is the composite of f^{-1} with the inverse $\text{dil}_{c^{-1}}$ of dil_c. Because similarities preserve the ratio of any two lengths, they preserve all angles. A very remarkable theorem shows, conversely, that similarities are the *only* maps of $\mathbb{R} \times \mathbb{R}$ that preserve all angles. The only proofs of this theorem I know of use complex analysis, which is well beyond the scope of this book. See, for example, Jones and Singerman (1987), p.200.

There is an equally remarkable theorem about the non-Euclidean plane that says that its only angle-preserving maps are the isometries. This means, in particular, that there are *no* maps of the non-Euclidean plane that multiply all lengths by a constant $\neq 1$. Hence beings in a non-Euclidean world would not enjoy the benefits of scale models and maps. On the other hand, they would be able to determine the size of a figure from its shape alone, because each shape (of a triangle, say) exists in only one size.

4 Rational Points

4.1 Pythagorean Triples

One of the most astonishing documents in the history of mathematics is a clay tablet in the Columbia University collection of Babylonian artifacts. Known as Plimpton 322, it dates from around 1800 B.C. and contains the two columns of numbers in Figure 4.1.

Few of the pairs (b, c) look at all familiar, and it is not obvious that they have any mathematical significance. However, they have a property that leaves no doubt what they are. *In every case, $c^2 - b^2$ is an integer square a^2*, hence the tablet is a virtual list of what we now call *Pythagorean triples* (a, b, c). In arithmetic terms, Pythagorean triples are simply triples of natural numbers with $a^2 + b^2 = c^2$, but by the converse Pythagorean theorem they are also *side lengths of right-angled triangles*. In fact, there is another column that shows the Babylonians were aware of this, and it explains why the pairs (b, c) are written in the given order. The column not included in the table here is a list of the values c^2/a^2, and they turn out to be in decreasing order, and roughly equally spaced. Thus the tablet is a kind of "database" of right-angled triangles, covering a range of shapes. Incidentally, it is an interesting question whether there are Pythagorean triples for which the angles of the corresponding triangles increase

111

b	c
119	169
3367	4825
4601	6649
12709	18541
65	97
319	481
2291	3541
799	1249
481	769
4961	8161
45	75
1679	2929
161	289
1771	3229
56	106

FIGURE 4.1   Pairs in Plimpton 322.

in *exactly* equal steps. We shall answer this question in Sections 5.4 and 5.8*.

The meaning of the pairs (b, c) was discovered by Otto Neugebauer and Abraham Sachs (1945), who also went on to speculate how the corresponding Pythagorean triples may have been found. This is a good question, because triples as large as $(13500, 12709, 18541)$ were certainly not found by trial and error, but the answer is probably not straightforward. The *purpose* of Plimpton 322 can be guessed by constructing the corresponding triangles, and the preference for certain numbers can be explained by the Babylonian number system (see Exercises 4.1.1 and 4.1.3), but for enlightenment on *method* we must turn to ancient Greece.

Exercises

4.1.1. Check that $c^2 - b^2$ is a perfect square for each of the pairs (b, c) in the table. (Computer assistance is recommended.)

Computing the values $a = \sqrt{c^2 - b^2}$ is interesting because in many cases a turns out to be a "rounder" number than b or c. In fact, our base

10 notation fails to show how very round the numbers a really are. All but three of them are multiples of 60 and, of the remaining three, one is a multiple of 30 and the others are multiples of 12. The Babylonians wrote their numbers in base 60, so multiples of 60 were the roundest numbers in their notation, and the divisors 12 and 30 of 60 were also pretty round.

4.1.2. What is significant about the number 3456?

It is also interesting to compute the values of $b/c = \sin\theta$, where θ is the angle opposite b in the right-angled triangle, and see how the angles increase in roughly equal steps.

4.1.3. Show that the values of b/c in Plimpton 322 strictly increase, and find the corresponding values of θ.

4.2 Pythagorean Triples in Euclid

The easiest way to find big Pythagorean triples is to use a formula like

$$a = 2uv, \quad b = u^2 - v^2, \quad c = u^2 + v^2.$$

This formula gives $a^2 + b^2 = c^2$ for any values of u and v, so by substituting natural numbers for u and v it is possible to obtain arbitrarily large triples. Formulas like this, or perhaps more special ones like

$$a = 2u, \quad b = u^2 - 1, \quad c = u^2 + 1,$$

were probably known in ancient Babylon, Greece, India, and China.

However, the first rigorous treatment of Pythagorean triples occurs in Euclid. He actually set out to solve a simpler problem: *to find two square numbers such that their sum is also a square (Elements, Book X, Lemma 1 to Proposition 28).* He started with (a geometric form of) the identity

$$xy + \left(\frac{x-y}{2}\right)^2 = \left(\frac{x+y}{2}\right)^2,$$

which he had established earlier (Book II, Proposition 5). He then observed that it is enough to choose x and y so that xy is a square, and that this is possible if x and y are "similar plane numbers". This means, in our language, that $x = u^2w$ and $y = v^2w$ for some natural numbers u, v, and w. (The reason for the name is that if a rectangle of area w is magnified by u or v, its area becomes u^2w or v^2w, respectively.) Substituting these "similar plane numbers" for x and y gives the identity

$$(uvw)^2 + \left(\frac{u^2 - v^2}{2}w\right)^2 = \left(\frac{u^2 + v^2}{2}w\right)^2,$$

hence Euclid has solved his problem. The numbers uvw, $\frac{u^2-v^2}{2}w$ and $\frac{u^2+v^2}{2}w$ he has found will be integers if u, v, and w are natural numbers and u, v are both odd or both even (the latter condition ensuring that 2 divides $u^2 - v^2$ and $u^2 + v^2$). Hence he has also found a formula to produce Pythagorean triples.

Euclid then made a throwaway remark that is even more interesting: *xy is a square only if x and y are similar plane numbers*. This is the key to finding all Pythagorean triples, because his numbers $\frac{x-y}{2} = b$, $\frac{x+y}{2} = c$ can equal *any* natural numbers $b < c$ by choosing $x = b + c$, $y = c - b$, and in this case $xy = c^2 - b^2$. If b and c belong to a Pythagorean triple, $c^2 - b^2 = xy$ must therefore be a square, and Euclid's remark is that this happens only if $x = u^2w$, $y = v^2w$ for some natural numbers u, v, w. Thus he is implicitly claiming the following result.

Parameterization of Pythagorean triples   *Any Pythagorean triple is of the form*

$$a = uvw, \quad b = \frac{u^2 - v^2}{2}w, \quad c = \frac{u^2 + v^2}{2}w$$

for some natural numbers u, v, and w.

Proof It remains to prove that xy is a square only if $x = u^2w$ and $y = v^2w$ for some natural numbers u, v, and w. In fact, Euclid did this in Proposition 2 of Book IX, which is based on his theory of divisibility. As mentioned in Section 1.6, it is equivalent, and often easier, to use unique prime factorization, and a proof along the latter lines goes as follows.

Suppose x and y are natural numbers and xy is a square. By removing $w = \gcd(x, y)$ from both x and y we obtain natural numbers

$x' = x/w$ and $y' = y/w$ for which $x'y'$ is also a square, but with $\gcd(x', y') = 1$. It follows that the unique prime factorizations of x' and y',

$$x' = p_1^{e_1} p_2^{e_2} \cdots p_r^{e_r},$$
$$y' = q_1^{f_1} q_2^{f_2} \cdots q_s^{f_s},$$

have no prime in common. But then, because

$$x'y' = p_1^{e_1} p_2^{e_2} \cdots p_r^{e_r} q_1^{f_1} q_2^{f_2} \cdots q_s^{f_s}$$

is a square, unique prime factorization implies that each of the exponents $e_1, e_2, \ldots, e_r, f_1, f_2, \ldots, f_s$ is even. That is, each exponent in the prime factorizations of x' and y' is even, and hence x' and y' are squares themselves. If $x' = u^2$ and $y' = v^2$ this gives $x = u^2 w$ and $y = v^2 w$, as required. $\qquad\square$

It is convenient to call numbers x', y' *relatively prime* when $\gcd(x', y') = 1$. The result that a product of relatively prime x', y' is a square only if x' and y' are squares is one of the most useful consequences of unique prime factorization. Similarly, one finds that the product of relatively prime numbers is a cube only if the numbers themselves are cubes, and so on.

Exercises

There are some variations on Euclid's formula for Pythagorean triples that are worth knowing. One is

$$a = 2uvw, \quad b = (u^2 - v^2)w, \quad c = (u^2 + v^2)w.$$

This is, of course, the double of Euclid's formula, and it does not look completely general, because $a = 2uvw$ is necessarily even. However, it is impossible for a and b both to be odd. If they were, a^2 and b^2 would both leave remainder 1 on division by 4, and hence $c^2 = a^2 + b^2$ would leave remainder 2. But c^2 is an even square and hence leaves remainder 0 on division by 4 (compare with the exercises to Section 1.2).

4.2.1. Deduce from these remarks that in any Pythagorean triple (a, b, c), if the sides are suitably ordered, a is even and b and c are either both even or both odd.

4.2.2. Now use the identity $4xy + (x - y)^2 = (x + y)^2$ to show that any pair $(b, c) = (x - y, x + y)$ of numbers that are both even or both odd extends to a Pythagorean triple (a, b, c) just in case $4xy$ is a square.

4.2.3. Use unique prime factorization to show that $4xy$ is a square if and only if $x = u^2 w$ and $y = v^2 w$ for some natural numbers u, v, w.

4.2.4. Deduce from the preceding exercises that any Pythagorean triple, if the sides are suitably ordered, is of the form

$$a = 2uvw, \quad b = (u^2 - v^2)w, \quad c = (u^2 + v^2)w$$

for some natural numbers u, v, w.

Pythagorean triples for which a, b, and c have no common divisor except 1 are called *primitive*.

4.2.5. Deduce from Exercise 4.2.4 that each primitive Pythagorean triple, suitably ordered, is of the form

$$a = 2uv, \quad b = u^2 - v^2, \quad c = u^2 + v^2,$$

where u and v are natural numbers with $\gcd(u, v) = 1$, one of them even and the other odd.

An interesting interpretation of the parameters u and v was given in the *Nine Chapters of Mathematical Art*, a Chinese work from the period between 200 B.C. and 200 A.D.

Suppose that one person walks along the sides a, b of a right-angled triangle at speed u, while another walks along the hypotenuse c at speed v, and that both cover the distance in the same time.

4.2.6. Show that, with a suitable choice of unit length,

$$a = 2uv, \quad b = u^2 - v^2, \quad c = u^2 + v^2.$$

(*Hint*: Use the speed condition to find an expression for $b + c$, and substitute it in $a^2 = c^2 - b^2 = (c - b)(c + b)$ to find an expression for $c - b$.)

4.3   Pythagorean Triples in Diophantus

Pythagorean triples may be grouped into classes in which each member of a class is an integer multiple of the smallest member. The smallest member (a, b, c) of each class is one for which a, b, and c have no common divisor > 1, or what we called a *primitive* Pythagorean triple in the exercises. From this viewpoint, we see that the main problem in finding Pythagorean triples is to find the primitive triples. Once we know $(3, 4, 5)$ is a Pythagorean triple, for example, it is trivial to list its multiples $(6, 8, 10), (9, 12, 15)\ldots$. In fact, they may all be regarded as the "same" triangle, with different choices of the unit of length.

The many integer triples (a, b, c) that are really the "same" may be condensed to a single *rational triple* $(a/c, b/c, 1)$, because if (a, b, c) and (a', b', c') are multiples of the same triple then $a/c = a'/c'$ and $b/c = b'/c'$. Rational numbers really simplify the story here, because we can find a formula for all rational Pythagorean triples without using unique prime factorization (or Euclid's equivalent theory of divisibility). This was discovered around 250 A.D. by the Greek mathematician Diophantus and presented in his book the *Arithmetica*.

Parameterization of rational Pythagorean triples *The nonzero rationals x and y such that $x^2 + y^2 = 1$ are the pairs of the form*

$$x = \frac{1 - t^2}{1 + t^2}, \qquad y = \frac{2t}{1 + t^2},$$

for rational numbers $t \neq 0, \pm 1$.

Proof The problem is to find points on the unit circle $x^2 + y^2 = 1$ with rational coordinates x and y, the so-called *rational points*. Some rational points are obvious, for example, $(-1, 0)$. We also notice that if (x_0, y_0) is any rational point, the line between it and $(-1, 0)$ has rational slope, namely, $t = y_0/(x_0 + 1)$.

Conversely, if $y = t(x+1)$ is any line through $(-1, 0)$ with rational slope t, then its second intersection with the unit circle is a rational point, as the following calculation shows. The intersection of $y = t(x + 1)$ with $x^2 + y^2 = 1$ occurs where

$$x^2 + t^2(x + 1)^2 = 1 \quad \text{(substituting } t(x + 1) \text{ for } y\text{)},$$

and hence

$$x^2(1 + t^2) + 2t^2x + t^2 - 1 = 0,$$

which has solutions

$$x = \frac{-2t^2 \pm \sqrt{4t^4 - 4(1 + t^2)(t^2 - 1)}}{2(1 + t^2)} \quad \text{by the quadratic formula}$$

$$= \frac{-t^2 \pm 1}{1 + t^2}$$

$$= -1, \frac{1 - t^2}{1 + t^2}.$$

The solution $x = -1$ gives the point $(-1, 0)$ we already know. The solution $x = (1 - t^2)/(1 + t^2)$ gives the second intersection (Figure 4.2), where

$$y = t(x + 1) = t\left(\frac{1 - t^2}{1 + t^2} + 1\right) = t\left(\frac{1 - t^2 + 1 + t^2}{1 + t^2}\right) = \frac{2t}{1 + t^2}.$$

Thus the coordinates of the second intersection are $x = \frac{1-t^2}{1+t^2}$, $y = \frac{2t}{1+t^2}$, and these are rational because they are built from the rational number t by rational operations.

Hence we have found all rational points on the circle, and all except $(-1, 0)$ have the form $\left(\frac{1-t^2}{1+t^2}, \frac{2t}{1+t^2}\right)$. By taking $t \neq 0, \pm 1$ we also exclude the points $(1, 0)$, $(0, 1)$ and $(0, -1)$. The formula $x = \frac{1-t^2}{1+t^2}$, $y = \frac{2t}{1+t^2}$ then covers exactly the rational Pythagorean triples, because the latter have x and y nonzero.   □

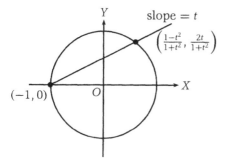

FIGURE 4.2   Constructing rational points on the circle.

Exercises

The formula for rational Pythagorean triples is appealing, because it involves only the single parameter t instead of the u, v, and w required for integer triples. Nevertheless, it is essentially the same formula.

4.3.1. By putting $t = u/v$ for integers u and v, deduce a formula for integer Pythagorean triples from the formula for rational Pythagorean triples.

The formula can in fact be simplified even further by using complex numbers, which we shall say more about later. Those already familiar with $\sqrt{-1} = i$ may enjoy the following exercise.

4.3.2. If points (x, y) of the plane are represented by complex numbers $x + iy$, show that the rational points on the unit circle, other than $(-1, 0)$, are the points of the form $\frac{t-i}{t+i}$, where t is rational.

The formulas

$$x = \frac{1 - t^2}{1 + t^2}, \qquad y = \frac{2t}{1 + t^2}$$

are also useful when t runs through all real values. The point (x, y) then runs through all points of the unit circle (except for $(-1, 0)$), hence the formulas may also be viewed as *parametric equations for the circle*. They are related to the more familiar parametric equations $x = \cos\theta$, $y = \sin\theta$, as we shall see in the next chapter.

We call the functions $x(t) = \frac{1-t^2}{1+t^2}$, $y(t) = \frac{2t}{1+t^2}$ *rational functions* of t because they are built from the variable t and constants by rational operations.

4.3.3. Find rational functions $x(t)$ and $y(t)$ such that $(x(t), y(t))$ runs through all points of the circle except $(1, 0)$.

Rational functions are simpler than the functions $\cos\theta$, $\sin\theta$, and they have certain advantages. Because of this, in algebra and calculus we often want to *rationalize* irrational functions $f(x)$ such as $\sqrt{1 - x^2}$. That is, we want to substitute a new function $x(t)$ for x so that $f(x)$ becomes a rational function of the new variable t.

4.3.4. Show that the function $\sqrt{1 - x^2}$ is rationalized by the substitution $x = \frac{1-t^2}{1+t^2}$, and also by the substitution $x = \frac{2t}{1+t^2}$.

In contrast to this result, the functions $\sqrt{1-x^3}$ and $\sqrt{1-x^4}$ can *not* be rationalized by substituting a rational function for x. This discovery marks an important boundary between quadratic and higher-degree polynomials and hence leads beyond the scope of this book. Nevertheless, properties of quadratic equations can be used to explain some of the difficulties that arise with higher degree, and we shall show how this comes about in Section 4.7*.

4.4　Rational Triangles

After the discovery of rational right-angled triangles and their complete description by Euclid (Section 4.2), one might expect questions to arise about rational triangles in general. Of course, any three rational numbers can be the sides of a triangle, provided the sum of any two of them is greater than the third. Thus a "rational triangle" should be one that is rational not only in its side lengths, but also in some other quantity, such as altitude or area. Because area $= \frac{1}{2}$base × height, a triangle with rational sides has rational area if and only all its altitudes are rational, so it is reasonable to define a *rational triangle* to be one with rational sides and rational area.

Many questions can be raised about rational triangles, but they rarely occur in Greek mathematics. As far as we know, the first to treat them thoroughly was the Indian mathematician Brahmagupta, in his *Brâhma-sphuṭa-siddhânta* of 628 A.D. In particular, he found the following complete description of rational triangles.

Parameterization of rational triangles　*A triangle with rational sides a, b, c and rational area is of the form*

$$a = \frac{u^2}{v} + v, \quad b = \frac{u^2}{w} + w, \quad c = \frac{u^2}{v} - v + \frac{u^2}{w} - w$$

for some rational numbers u, v, and w.

Brahmagupta (see Colebrooke (1817), p. 306) actually has a factor $1/2$ in each of a, b, and c, but this is superfluous because, for example,

$$\frac{1}{2}\left(\frac{u^2}{v} + v\right) = \frac{(u/2)^2}{v/2} + v/2 = \frac{u_1^2}{v_1} + v_1,$$

where $u_1 = u/2$ and $v_1 = v/2$ are likewise rational. The formula is stated without proof, but it becomes easy to see if one rewrites a, b, c and makes the following stronger claim.

Any triangle with rational sides and rational area is of the form

$$a = \frac{u^2 + v^2}{v}, \quad b = \frac{u^2 + w^2}{w}, \quad c = \frac{u^2 - v^2}{v} + \frac{u^2 - w^2}{w}$$

for some rationals u, v, and w, with altitude $h = 2u$ splitting side c into segments $c_1 = \frac{u^2 - v^2}{v}$ and $c_2 = \frac{u^2 - w^2}{w}$.

The stronger claim says in particular that any rational triangle splits into two rational right-angled triangles. It follows from the parameterization of rational right-angled triangles and was presumably known to Brahmagupta.

Proof For *any* triangle with rational sides a, b, c, the altitude h splits c into rational segments c_1 and c_2 (Figure 4.3). This follows from the Pythagorean theorem in the two right-angled triangles with sides c_1, h, a and c_2, h, b respectively, namely,

$$a^2 = c_1^2 + h^2,$$
$$b^2 = c_2^2 + h^2.$$

Hence, by subtraction,

$$a^2 - b^2 = c_1^2 - c_2^2 = (c_1 - c_2)(c_1 + c_2) = (c_1 - c_2)c,$$

so

$$c_1 - c_2 = \frac{a^2 - b^2}{c}, \quad \text{which is rational.}$$

But also

$$c_1 + c_2 = c, \quad \text{which is rational,}$$

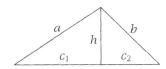

FIGURE 4.3 Splitting a rational triangle.

hence

$$c_1 = \frac{1}{2}\left(\frac{a^2 - b^2}{c} + c\right), \quad c_2 = \frac{1}{2}\left(c - \frac{a^2 - b^2}{c}\right)$$

are both rational.

Thus if the area, and hence the altitude h, are also rational, the triangle splits into two rational right-angled triangles with sides c_1, h, a and c_2, h, b.

We know from Diophantus' method (4.3) that any rational right-angled triangle with hypotenuse 1 has sides of the form

$$\frac{1 - t^2}{1 + t^2}, \quad \frac{2t}{1 + t^2}, \quad 1 \qquad \text{for some rational } t,$$

or, writing $t = v/u$,

$$\frac{u^2 - v^2}{u^2 + v^2}, \quad \frac{2uv}{u^2 + v^2}, \quad 1 \qquad \text{for some rational } u, v.$$

Thus the arbitrary rational right-angled triangle with hypotenuse 1 is a multiple (by $\frac{v}{u^2+v^2}$) of the triangle with sides

$$\frac{u^2 - v^2}{v}, \quad 2u, \quad \frac{u^2 + v^2}{v}.$$

The latter therefore represents all rational right-angled triangles with altitude $2u$, as the rational v varies. It follows that any *two* rational right-angled triangles with altitude $2u$ have sides

$$\frac{u^2 - v^2}{v}, \quad 2u, \quad \frac{u^2 + v^2}{v} \qquad \text{and} \qquad \frac{u^2 - w^2}{w}, \quad 2u, \quad \frac{u^2 + w^2}{w}$$

for some rational v and w. Putting the two together (Figure 4.4) gives an arbitrary rational triangle, and its sides and altitude are of the required form.   □

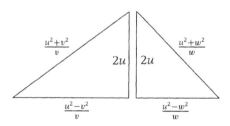

FIGURE 4.4   Assembling an arbitrary rational triangle.

Exercises

4.4.1. (Brahmagupta) Show that the triangle with sides 13, 14, 15 splits into two integer right-angled triangles.

Triangles with rational sides and rational area are sometimes called *Heronian* after the Greek mathematician Hero who lived in the first century A.D. Hero is also known for a formula giving the area of a triangle in terms of the lengths of its sides. His formula can, in fact, be derived quite easily from Brahmagupta's formulas for the sides of a rational triangle.

4.4.2. Show that for *any* triangle with sides a, b, c and altitude h on side c there *real* numbers u, v, w such that

$$a = \frac{u^2 + v^2}{v}, \quad b = \frac{u^2 + w^2}{w}, \quad c = \frac{u^2 - v^2}{v} + \frac{u^2 - w^2}{w},$$

with the side c split into parts $\frac{u^2-v^2}{v}$ and $\frac{u^2-w^2}{w}$ by the altitude $h = 2u$.

4.4.3. Define the *semiperimeter* s of the triangle with sides a, b, and c to be $(a + b + c)/2$. Then, with the notation of Exercise 4.4.2, show that

$$s(s - a)(s - b)(s - c) = u^2(v + w)^2 \left(\frac{u^2}{vw} - 1 \right)^2.$$

4.4.4. Deduce from Exercise 4.4.3 that

$$\sqrt{s(s - a)(s - b)(s - c)} = u \left(\frac{u^2 - v^2}{v} + \frac{u^2 - w^2}{w} \right)$$

is the area of the triangle with sides a, b and c.

Area $= \sqrt{s(s - a)(s - b)(s - c)}$ is Hero's formula. It defies the Greek geometric tradition by multiplying four lengths—something usually rejected as physically meaningless. Brahmagupta probably was aware of this formula, because he stated a generalization of it: the area of a *cyclic quadrilateral* (a four-sided polygon with its vertices on a circle) is $\sqrt{(s - a)(s - b)(s - c)(s - d)}$, where a, b, c, d are the sides of the quadrilateral and s is half its perimeter.

4.5   Rational Points on Quadratic Curves

The method for finding rational Pythagorean triples can also be used to find rational points on other quadratic curves. If we know one rational point P, then any other rational point Q will give a line PQ with rational slope, hence all rational points occur on lines through P with rational slope. Conversely, a line with rational slope, through one rational point, will meet the curve in a second rational point, *provided the coefficients in the equation of the curve are all rational.*

To see that rational coefficients are needed, consider the curve $y = \sqrt{2}x^2$. This has one rational point $(0, 0)$, and the line $y = x$ through this point has slope 1. But the line meets the curve again where $x = \sqrt{2}x^2$, that is, at the irrational point $x = 1/\sqrt{2}$.

To see what happens when the coefficients are rational, consider the curve $x^2 + 3y^2 = 1$. This has an obvious rational point $(-1, 0)$, and if we take the line $y = t(x + 1)$ through this point with rational slope t, its intersections with the curve are found by substituting $t(x + 1)$ for y in the equation for the curve. This gives the quadratic equation in x,

$$x^2 + 3t^2(x + 1)^2 - 1 = 0.$$

which we will *not* attempt to solve this time (though it is not hard). Instead, bear in mind that $x = -1$ is a solution of this equation, and hence $x + 1$ is a factor of the left-hand side. If we expand the left-hand side, we shall find a rational coefficient k of x^2 (built from the rational t by $+$, $-$, and $\times$), and therefore

$$x^2 + 3t^2(x + 1)^2 - 1 = k(x + 1)(x - u)$$

where $x = u$ is the other solution of the equation. It follows, by comparing coefficients on the two sides, that ku is the negative of the constant term on the left-hand side. This constant term is also rational, because it is built from the rational t by $+$, $-$ and $\times$. Thus the x coefficient u of the second point of intersection is rational, and hence so is the y coefficient, because $y = t(x + 1)$.

Similar reasoning applies to any quadratic equation with rational coefficients, hence we have the following.

Description of the rational points on a quadratic curve *If a curve $\mathcal{K}$ is given by a quadratic equation with rational coefficients, then the rational points on $\mathcal{K}$ consist of*

1. *Any single rational point P on $\mathcal{K}$.*
2. *The points where lines through P with rational slope meet $\mathcal{K}$.*

It is not claimed that a curve with rational coefficients has *any* rational points. However, if it has one, it has infinitely many, because there are infinitely many lines of rational slope through any point P.

Example *The curve $x^2 + y^2 = 3$ has no rational points.*

First note that any rational point (x, y) has $x = u/w$, $y = v/w$ for some integers u, v, and w (with w the common denominator of x and y). It follows, multiplying through by w^2, that a rational point on $x^2 + y^2 = 3$ gives integers satisfying

$$u^2 + v^2 = 3w^2.$$

We can assume that u, v, w have no common divisor > 1, so they are not all even. Then at least one of u and v is odd, because if u, v are even so is $u^2 + v^2 = 3w^2$, and $3w^2$ is even only if w is even. However ...

1. If u, v are both odd then u^2, v^2 both leave remainder 1 on division by 4, hence $u^2 + v^2$ leaves remainder 2 (compare with the exercises to Section 4.2). But $3w^2$ leaves remainder 3 (if w is odd) or 0 (if w is even).
2. If one of u, v is odd and the other even, then $u^2 + v^2$ leaves remainder 1 on division by 4, which again is not the remainder left by $3w^2$.

Thus, in all cases an integer solution of $u^2 + v^2 = 3w^2$ gives a contradiction, hence there is no rational point on $x^2 + y^2 = 3$. $\square$

Probably the first result of this type was discovered by Diophantus, who stated that $x^2 + y^2 = 15$ has no rational solution (*Arithmetica* Book VI, Problem 14; see Heath (1910), p.237). The argument for $x^2 + y^2 = 15$ is virtually the same as the argument for $x^2 + y^2 = 3$, because $15w^2$ leaves the same remainder on division by 4 as $3w^2$ does.

These examples remind us that questions about rational numbers are basically questions about integers, and sometimes we have to go back to the integers to answer them. Nevertheless, rational points on curves are generally easier to find than integer points. This is already clear for the line $ax + by = c$ with integer coefficients, where integer points exist only when $\gcd(a, b)$ divides c, and finding them amounts to finding the gcd (see Section 1.5). Rational points always exist, and we can find them simply by solving for $y = (c - ax)/b$ and letting x run through the rationals. (Or, if $b = 0$, solve for x in terms of y.) Deeper problems occur with the quadratic curves $x^2 - dy^2 = 1$. Their rational points are no harder to find than rational points on the unit circle, but finding integer points is an entirely different matter (see Chapters 8 and 9).

Exercises

The curves $x^2 - dy^2 = 1$ for $d > 0$ are called *hyperbolas* and some of their geometric properties will be studied in Chapters 8 and 9. The geometry has some bearing on the behavior of integer points, as we shall see in Chapter 9, but algebra also plays an important role, as we shall see in Chapter 8. For the moment, we shall investigate these curves as best we can with our current tools.

4.5.1. Show that the hyperbola $x^2 - dy^2 = 1$ approaches arbitrarily close to the lines $x = \pm\sqrt{d}y$, and hence sketch the curve.

4.5.2. Show that the rational points other than $(-1, 0)$ on $x^2 - dy^2 = 1$ are given by the formulas

$$x = \frac{1 + dt^2}{1 - dt^2}, \qquad y = \frac{2t}{1 - dt^2}.$$

You will probably find that these formulas are no help in finding integer points on $x^2 - dy^2 = 1$, other than the obvious ones $(-1, 0)$ and $(1, 0)$. In fact, the integer points depend mysteriously on the value of d, which we assume to be a natural number from now on.

4.5.3. By factorizing the left-hand side, show that there are no integer points on the hyperbola $x^2 - y^2 = 1$, other than the obvious ones.

4.5.4. Find a nonobvious integer point on each of the hyperbolas $x^2 - 2y^2 = 1$, $x^2 - 3y^2 = 1$, and $x^2 - 5y^2 = 1$. What happens on $x^2 - 4y^2 = 1$?

One expects that rational points on curves with irrational coefficients are not so interesting, because they presumably occur only "by accident." But if so, it is still worth making this presumption more precise. In fact they *are* accidental, in the sense that there are only finitely many rational points on each curve with irrational coefficients. We shall confine attention to quadratic curves and assume further that their equations are of the form $ax^2 + by^2 = c$ with c rational. (This can always be arranged by shift of origin and rotation of axes, as we shall see in Chapter 8.)

4.5.5. Suppose that $\mathcal{K}$ is a curve given by $ax^2 + by^2 = c$, and that $\mathcal{K}$ has infinitely many rational points. Deduce that the coefficients a, b satisfy infinitely many equations of the form

$$Aa + Bb = c, \quad \text{for rational numbers } A, B.$$

4.5.6. Deduce from Exercise 4.5.5 that $\mathcal{K}$ has rational coefficients.

Even when a quadratic curve $\mathcal{K}$ has only finitely many rational points, the idea of considering the line of slope t through a point on $\mathcal{K}$ is fruitful, because it shows that x and y can always be expressed as rational *functions* of t. (Recall from the comment after Exercise 4.3.2 that a rational function of t is built from t and constants by rational operations. The constants need not be rational.)

4.5.7. If $\mathcal{K}$ is the curve $ax^2 + bxy + cy^2 + dx + ey + f = 0$, and (r, s) is any point on $\mathcal{K}$, use the line through (r, s) with slope t to find parametric equations for $\mathcal{K}$, $x = u(t)$, $y = v(t)$, where $u(t)$ and $v(t)$ are rational functions of t.

4.6* Rational Points on the Sphere

The geometric construction used in Section 4.2 to find rational points on the circle can be viewed as *projection* of the y-axis onto the circle minus the point $(-1, 0)$. In fact, it is precisely the point $y = t$ that is projected to the point $\left(\frac{1-t^2}{1+t^2}, \frac{2t}{1+t^2}\right)$ (Figure 4.5).

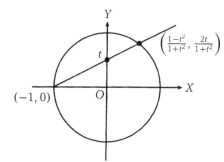

FIGURE 4.5 Projection from line to circle.

Projection from line to circle has a generalization, called *stereographic projection*, from plane to sphere. The (x, y)-plane in (x, y, z)-space is mapped to the unit sphere $x^2 + y^2 + z^2 = 1$ by projection toward the "north pole" $N = (0, 0, 1)$ (Figure 4.6). (Strictly speaking, stereographic projection goes from sphere to plane, but we are interested in both directions.)

Formulas for stereographic projection　*If* $P = (u, v)$ *in the plane and* $P' = (p, q, r)$ *on the sphere correspond under stereographic projection, then*

$$u = \frac{p}{1 - r}, \qquad v = \frac{q}{1 - r}$$

and

$$p = \frac{2u}{u^2 + v^2 + 1}, \qquad q = \frac{2v}{u^2 + v^2 + 1}, \qquad r = \frac{u^2 + v^2 - 1}{u^2 + v^2 + 1}.$$

Proof　The line through $N = (0, 0, 1)$ and $P' = (p, q, r)$ has the direction components $p, q, r - 1$, and hence parametric equations

$$x = pt, \quad y = qt, \quad z = 1 + (r - 1)t.$$

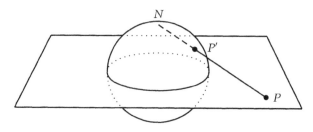

FIGURE 4.6
Projection from plane to sphere.

It meets the (x, y)-plane at P where $z = 0$, that is, where $t = \frac{1}{1-r}$, so $\frac{p}{1-r}$, $\frac{q}{1-r}$ are the coordinates u, v of P.

The line through $N = (0, 0, 1)$ and $P = (u, v)$ has direction components u, v, -1; hence parametric equations

$$x = ut, \qquad y = vt, \qquad z = 1 - t.$$

Substituting these in the equation $x^2 + y^2 + z^2 = 1$ of the sphere, we get the equation

$$u^2 t^2 + v^2 t^2 + (1 - t)^2 = 1$$

for the parameter value t at the intersection. This equation simplifies to

$$t^2(u^2 + v^2 + 1) - 2t = 0.$$

One solution $t = 0$ corresponds to N. P' corresponds to the other solution

$$t = \frac{2}{u^2 + v^2 + 1},$$

which gives

$$x = \frac{2u}{u^2 + v^2 + 1}, \qquad y = \frac{2v}{u^2 + v^2 + 1},$$

$$z = 1 - \frac{2}{u^2 + v^2 + 1} = \frac{u^2 + v^2 - 1}{u^2 + v^2 + 1}$$

as the coordinates p, q, r of P'. $\qquad\qquad\qquad\qquad\square$

The formulas show that p, q, and r are rational if and only if u and v are rational. Hence we have the following.

Corollary *The rational points $(p, q, r) \neq (0, 0, 1)$ on the unit sphere are*

$$p = \frac{2u}{u^2 + v^2 + 1}, \qquad q = \frac{2v}{u^2 + v^2 + 1}, \qquad r = \frac{u^2 + v^2 - 1}{u^2 + v^2 + 1}$$

for rational u and v.

The idea of stereographic projection applies to space of any dimension n, though naturally it is difficult to visualize when $n > 3$, and the formulas take over. However, from the two cases we know, it is easy to see what to do next. The n-dimensional unit sphere in $(x_1, x_1, \dots, x_n)$-space has equation

$$x_1^2 + x_2^2 + \cdots + x_n^2 = 1,$$

and its rational points $(p_1, p_2, \dots, p_n)$ are found by connecting the "north pole" $(0, 0, \dots, 0, 1)$ to the point $(u_1, u_2, \dots, u_{n-1}, 0)$ for rational values of $u_1, u_2, \dots, u_{n-1}$. The coordinates $p_1, p_2, \dots p_n$ turn out to be

$$p_1 = \frac{2u_1}{u_1^2 + u_2^2 + \cdots + u_{n-1}^2 + 1}, \quad \cdots, \quad p_n = \frac{u_1^2 + u_2^2 + \cdots + u_{n-1}^2 - 1}{u_1^2 + u_2^2 + \cdots + u_{n-1}^2 + 1}.$$

Exercises

Each rational point $(\frac{a}{d}, \frac{b}{d}, \frac{c}{d})$ on the sphere $x^2 + y^2 + z^2 = 1$ corresponds to an integer quadruple (a, b, c, d) such that

$$a^2 + b^2 + c^2 = d^2,$$

so the formulas give a way to find all such quadruples.

4.6.1. Find formulas that give all such quadruples (a, b, c, d).

4.6.2. Do your formulas give the quadruples $(1, 2, 2, 3)$ and $(1, 4, 8, 9)$?

The projection of the plane onto the sphere minus N generalizes to any surface given by a quadratic equation in x, y, z.

4.6.3. Find a rational point T on the surface $2x^2 + 3y^2 + 4z^2 = 5$, and hence find formulas for all its rational points $\neq T$.

4.6.4.* If S is a surface given by a quadratic equation with rational coefficients, show that the rational points on S (if any) may be obtained by projecting from any rational point T off the (x, y)-plane to the rational points on the (x, y)-plane.

Thus, as with curves, a quadratic surface with rational coefficients has either no rational points or infinitely many.

4.6.5.* Show that the sphere $x^2 + y^2 + z^2 = 7$ has no rational points. (*Hint*: Consider remainders on division by 8.)

4.7* The Area of Rational Right Triangles

In this section we return to the interpretation of a rational Pythagorean triple (a, b, c) as a right-angled triangle with rational sides a, b, c: we shall call it the *rational right triangle* (a, b, c). Geometry suggests some interesting questions about such a triangle (a, b, c). For example, what can we say about its area? Diophantus answered many questions of this type in Book VI of his *Arithmetica*. He found triangles (a, b, c) whose area $ab/2$ is a square $\pm$ a given number, a square $\pm$ the sum of the perpendiculars, a square minus the hypotenuse, and a square minus the perimeter. However, the possibility of the area being *exactly* a square is ignored!

The first to ask whether there is such a triangle was Fibonacci, who raised the question in his *Liber Quadratorum* (book of squares) in 1225. (Strictly speaking, he asked an equivalent question; see Sigler (1987) p. 84.) In 1640 Fermat proved that the answer is no. His proof is a spectacular application of infinite descent to Pythagorean triples, and several variations of it exist. The following version is based on Young (1992). It assumes the formula for primitive Pythagorean triples from the exercises to Section 4.2.

Fermat's theorem on rational right triangles *The area of a rational right triangle is not a square.*

Proof Given any rational right triangle, we can take its sides to be integers with no common prime divisor, by multiplying through by a common denominator and canceling any common prime factors. This process multiplies its area by a square (because the base and height are both multiplied by the same factor), so if there is a rational right triangle with square area, there is a primitive Pythagorean triple (a, b, c) with $ab/2$ a square. The strategy of the proof is to look for a smaller triangle with the same property.

The formula for primitive Pythagorean triples gives natural numbers u and v such that

$$a = 2uv, \quad b = u^2 - v^2, \quad c = u^2 + v^2,$$

where $\gcd(u, v) = 1$ and one of u, v is even, the other odd. It follows that the area of triangle (a, b, c) is

$$\frac{ab}{2} = uv(u^2 - v^2) = uv(u - v)(u + v).$$

The factors u, v, $u - v$, $u + v$ have no common prime divisor, as one checks by comparing them in pairs. A common prime divisor of u and $u - v$ also divides their difference, v, and we know that u and v have no such divisor. Similarly, the pairs u, $u + v$ and v, $u - v$ and v, $u + v$ each have no common prime divisor. Finally, a common prime divisor of $u - v$ and $u + v$ divides their sum $2u$ and their difference $2v$. Because one of u, v is even and the other is odd, $u - v$ is odd and hence 2 is *not* a divisor of $u - v$, $2u$ and $2v$. Any common prime divisor must then divide u and v, and hence it does not exist.

Thus a square area $ab/2 = uv(u - v)(u + v)$ has factors u, v, $u - v$, and $u + v$, which are themselves squares, by unique prime factorization. It follows that $u^2 - v^2 = (u - v)(u + v)$ is a product of squares, hence also a square, say w^2. This gives us

$$u^2 - v^2 = w^2, \quad \text{or} \quad v^2 + w^2 = u^2,$$

so (v, w, u) is a *second Pythagorean triple*. We already know $\gcd(u, v) = 1$, so the new triple is primitive, hence there are natural numbers u_1 and v_1 with

$$v = 2u_1 v_1, \quad w = u_1^2 - v_1^2, \quad u = u_1^2 + v_1^2.$$

(We know that v is the even member $2u_1 v_1$ because $w^2 = u^2 - v^2$ is odd, hence w is the odd member.)

Because $u = u_1^2 + v_1^2$ is a square, say w_1^2, we have a *third Pythagorean triple* (u_1, v_1, w_1). The area of the corresponding right triangle is $u_1 v_1/2 = v/4$, which is a square (because v is a square, as we found in the previous paragraph). Thus we have found another triangle with the same property as the first.

The third triangle still has natural number sides and natural number area, but its area $v/4$ is less than the area $uv(u-v)(u+v)$ of the first triangle. Therefore, if there is a rational right triangle with square area, we can make an infinite descent, which is impossible.   □

Fermat drew some conclusions from this argument, which are as remarkable as the theorem itself.

Corollaries

1. *There are no natural numbers a, b, c such that $a^4 - b^4 = c^2$.*

2. *There are no natural numbers x, y, z such that $x^4 + y^4 = z^4$.*

Proof 1. For any natural numbers a, b, and c, consider the triangle with sides

$$a^4 - b^4, \quad 2a^2b^2, \quad a^4 + b^4.$$

This is a right-angled triangle because

$$(a^4 - b^4)^2 + (2a^2b^2)^2 = (a^4 + b^4)^2.$$

But if $a^4 - b^4 = c^2$, its area $(a^4 - b^4)a^2b^2$ is the square $a^2b^2c^2$, which contradicts the theorem. Hence $a^4 - b^4 = c^2$ is impossible for natural numbers a, b, and c.

2. If $x^4 + y^4 = z^4$ for natural numbers x, y, and z then

$$z^4 - y^4 = x^4 = (x^2)^2,$$

which is a special case of the equation proved impossible in part 1. Hence there are no such natural numbers x, y, and z.   □

Fermat also proved the impossibility of the equation $a^4 + b^4 = c^2$ in the natural numbers. A proof is outlined in the exercises.

Exercises

The structure of the proof of Fermat's theorem on rational right triangles can be presented quite concisely if the checks on divisibility are left to the reader. It goes as follows:

(a, b, c) a primitive Pythagorean triple with $ab/2$ a square

$\Rightarrow a = 2uv, \quad b = u^2 - v^2, \quad c = u^2 + v^2$ with $uv(u^2 - v^2)$ a square,

for some natural numbers u and v

$\Rightarrow u, v, u - v, u + v$ are squares

$\Rightarrow u^2 - v^2 = (u - v)(u + v) = w^2$ for some natural number w

$\Rightarrow (v, w, u)$ a primitive Pythagorean triple

$\Rightarrow v = 2u_1 v_1, \quad w = u_1^2 - v_1^2, \quad u = u_1^2 + v_1^2,$

for some natural numbers u_1 and v_1

$\Rightarrow u_1^2 + v_1^2$ is a square, say w_1^2, because u is a square

$\Rightarrow (u_1, v_1, w_1)$ a Pythagorean triple, with $u_1 v_1/2 = v/4$ a square

$\Rightarrow$ infinite descent, because $v/4 < ab/2 = uv(u^2 - v^2)$

The impossibility of $a^4 + b^4 = c^2$ is usually proved with the help of a formula for Pythagorean triples, but this step can be bypassed. The following proof from Cassels (1991) uses more basic facts about remainders on division by 2 and 4, together with unique prime factorization. It begins by assuming that a, b, and c have no common prime divisor, and $a^4 + b^4 = c^2$. Then it follows that c is odd, and so is one of the others, say b, by the argument preceding Exercise 4.2.1.

4.7.1. Check the details in the following proof:

$$a^4 + b^4 = c^2, \quad \begin{array}{l} \text{with no common prime divisor of } a, b, c \\ \text{and } a \text{ even} \end{array}$$

$\Rightarrow (c + b^2)(c - b^2) = a^4$

$\Rightarrow c + b^2 = 8u^4$ and $c - b^2 = 2v^4$

$\quad$ or $c + b^2 = 2u^4$ and $c - b^2 = 8v^4$

$\Rightarrow b^2 = 4u^4 - v^4 \quad \begin{array}{l} \text{(impossible, considering remainders on} \\ \text{division by 4)} \end{array}$

$\quad$ or $b^2 = u^4 - 4v^4$

$\Rightarrow (u^2 + b)(u^2 - b) = 4v^4$

$\Rightarrow u^2 + b = 2r^4$ and $u^2 - b = 2s^4$

$\Rightarrow u^2 = r^4 + s^4$

$\Rightarrow$ infinite descent

In Section 4.3 it was pointed out that the formula for rational Pythagorean triples gives us functions that rationalize the irrational function $\sqrt{1 - x^2}$. For example, if we substitute $x = \frac{1-t^2}{1+t^2}$ we find $\sqrt{1 - x^2} = \frac{2t}{1+t^2}$. Fermat's results about fourth powers can be similarly used to prove that the functions $\sqrt{1 - x^4}$ and $\sqrt{1 + x^4}$ can *not* be rationalized. In the latter case, for example, the idea is to suppose that there is some rational function $x(t)$ such that $\sqrt{1 + x(t)^4}$ is a rational function $y(t)$ and derive a contradiction. A rational function is a quotient of polynomials, so we are supposing that there are polynomials $p(t)$, $q(t)$, $r(t)$, $s(t)$ such that

$$\sqrt{1 + \frac{p(t)^4}{q(t)^4}} = \frac{r(t)}{s(t)},$$

or equivalently,

$$s(t)^4 \left(q(t)^4 + p(t)^4\right) = q(t)^4 r(t)^2 s(t)^2.$$

This yields polynomials $a(t) = s(t)q(t)$, $b(t) = s(t)p(t)$, and $c(t) = q(t)^2 r(t)s(t)$ with

$$a(t)^4 + b(t)^4 = c(t)^2,$$

which is the same as the Fermat equation, but with polynomials in place of the natural numbers a, b, and c. It can be proved impossible by imitating the argument given earlier because polynomials behave a lot like natural numbers. The degree of a polynomial serves as measure of its size, which can be used in proofs by induction (or descent).

4.7.2.* Show that polynomials have the following *division property*. If $a(t)$ and $b(t)$ are polynomials and $b(t)$ has degree > 0, then

$$a(t) = q(t)b(t) + r(t)$$

for some polynomials $q(t)$ and $r(t)$, with $r(t)$ of smaller degree than $b(t)$.

The theory of divisibility and factorization now unfolds for polynomials just as it did for natural numbers in Sections 1.5 and 1.6. The polynomials analogous to primes are called *irreducibles*.

4.7.3.* Check that there is a Euclidean algorithm for polynomials, an irreducible divisor property, and unique factorization into irreducibles (up to the order of factors and constant multiples of factors).

4.7.4.* Deduce from Exercise 4.7.3* that if the product of relatively prime polynomials is a square, then each factor is itself a square.

We can now imitate the argument of Exercise 4.7.1 with polynomials in place of natural numbers, but it is *easier* because polynomials need not have rational coefficients. If $p(t)$ and $q(t)$ are relatively prime polynomials and $p(t)q(t)$ is a square, we can conclude not only that $p(t) = u(t)^2$ and $q(t) = v(t)^2$, but also that $p(t) = 2U(t)^2$ and $q(t) = 2V(t)^2$, for the polynomials $U = u/\sqrt{2}$ and $V = v/\sqrt{2}$. This means it is no longer necessary to worry about the coefficients 2, 4, and 8.

4.7.5.* By imitating the argument in Exercise 4.7.1, show that there are
no polynomials $a(t)$, $b(t)$, $c(t)$ of degree > 0 such that

$$a(t)^4 + b(t)^4 = c(t)^2.$$

4.8 Discussion

Diophantus and His Legacy

The last peak in classical Greek mathematics was reached by Diophantus of Alexandria, sometime between 150 A.D. and 300 A.D. The
surviving parts of his work, the *Arithmetica*, seem at first quite elementary, a random collection of solved problems about numbers.
There are no general theorems, and there is no apparent "depth,"
because later results do not depend on earlier ones, as they do in
Euclid's *Elements*. However, this apparent simplicity is deceptive.
The problems of Diophantus effectively illustrate general theorems,
and some of them were deep enough to inspire Fermat and Euler,
the greatest number theorists of the 17th and 18th centuries. Euler
wrote:

> Diophantus himself, it is true, gives only the most special
> solutions of all the questions which he treats, and he is gen-
> erally content with indicating numbers which furnish one
> single solution. But it must not be supposed that his method
> is restricted to these very special solutions. In his time the use
> of letters to denote undetermined numbers was not yet estab-
> lished, and consequently the more general solutions which
> we are enabled to give by means of such notation could not be
> expected from him. Nevertheless, the actual methods which
> he uses for solving any of his problems are as general as
> those which are in use today; nay, we are obliged to admit
> that there is hardly any method yet invented in this kind
> of analysis of which there are not sufficiently distinct traces
> to be discovered in Diophantus. (Euler *Opera Omnia* 1, II,
> p.429–430, translated by Heath (1910) p. 56)

Diophantus' success can be partly explained by his innovations in notation, which enabled him to carry out more complex algebraic manipulations than his predecessors. He used a symbol for the unknown, and abbreviations for the arithmetic operations, which were sufficient to solve certain polynomial equations and compute with complicated fractions. The limitation of his notation is that there is only *one* symbol for an unknown, so problems with several unknowns are solved by choosing particular values for all but one of them. This is why he restricts himself to particular problems—and why the restriction is not severe, as Euler realized.

The distinctive feature of Diophantus' work is an interest in *rational* solutions of equations. In some ways, rational numbers are easier to work with than integers, so Diophantus had the advantage of being first into a field his predecessors were not equipped to explore. However, he brought to this field exceptional ingenuity and insight. His ideas were not fully understood, let alone extended, until Fermat reconsidered them in the 17th century.

Diophantus' subject matter is now called *Diophantine equations*, a rather misleading term that replaces the equally misleading "indeterminate equations" found in older books. It would be better described as *finding rational solutions of equations*. Typically, the equations considered have infinitely many solutions (hence the term *indeterminate*) and the challenge is to find the rational solutions, if any. Since the time of Fermat, it has been recognized that finding integer solutions is an even more challenging problem, and the term *Diophantine equations* is sometimes reserved for the subject with this narrower aim. Today, mathematicians have come to view these subjects geometrically, and they are often described as *finding rational points on curves* and *finding integer points on curves*. Is this really what Diophantus was doing? He did not say so, but his solutions are open to both algebraic and geometric interpretations.

The classic source of Diophantus in English is the translation and commentary by Heath (1910). This book is still the most complete and informative, and incidentally it's also a superb introduction to the number theory of Fermat and Euler. However, Heath views Diophantus purely as an algebraist, and to see the geometric side of the story, it is also advisable to read Weil (1984).

Several interesting problems are conspicuous by their absence from the *Arithmetica*. Diophantus sometimes skirts around a problem, answering several questions but not the one that seems most central. It looks like he has been stumped, then (like a student faced with a similar situation on an exam) decided to tell what he knows about something else. He is answering related questions, but with extra conditions that make them easier to solve. The missing questions were eventually raised by readers of the *Arithmetica*, particularly Fermat, and it became clear that new ideas were needed to answer them. Fermat claimed solutions, but divulged very few; most of the published solutions were by Euler and Lagrange. We shall study some of their innovations later, but it is appropriate to mention the questions here.

The first two arise from the study of rational right-angled triangles, as we saw in Section 4.7*.

1. *Can the area of a rational right-angled triangle be a square?*
   All of Book VI in the *Arithmetica* is concerned with rational right-angled triangles. As mentioned in Section 4.7*, Diophantus finds examples whose area is a square ± a given number, a square ± the sum of the perpendiculars, a square minus the hypotenuse, and a square minus the perimeter—virtually everything *except* a square.

   Fermat proved that the latter is impossible, by an argument similar to that given in Section 4.7*.

2. *Can the sum of two fourth powers be a square?*
   In Book V, Problem 29, Diophantus gave an example of three numbers, 144/25, 9 and 16, whose fourth powers sum to a square.

   Fermat asked about the sum of two fourth powers, and showed that the answer is no. It is remarkable that the answer comes from his proof that the area of a rational right-angled triangle is not a square.

3. *Is every positive integer the sum of four squares?*
   This question was raised in 1621 by Bachet, whose edition of Diophantus was the one used by Fermat. Bachet was prompted by Diophantus' Problem 29 of Book IV, which answers a more complicated question about sums of squares.

Fermat claimed he could prove that every positive integer is the sum of four squares, Euler attacked the problem without complete success, and the first solution was published by Joseph Louis Lagrange in 1770.

4. *Is every prime of the form* $4n + 1$ *the sum of two squares?*
This question arises from Problem 19 of Book III, which is discussed further in Section 7.1.

Fermat claimed a proof in 1640, but the first published proof was by Euler in 1749. It was followed by many other proofs, one of which is presented in Section 7.6.

5. *Is* $x = 5$, $y = 3$ *the only positive integer solution of* $y^3 = x^2 + 2$?
Diophantus gives this solution in Book VI, Problem 17, though without claiming it is unique.

Fermat in 1657 claimed that it was the only positive integer solution, and a remarkable proof was given by Euler (1770). Euler's proof (with some necessary slight corrections) is in the exercises to Section 7.6.

Fermat's Last Theorem

The formula for rational Pythagorean triples credited to Diophantus in Section 4.3 is not exactly what he wrote. However, this formula can be read between the lines of Problem 8 in Book II of the *Arithmetica*: splitting a square into two squares. Because this problem is important for other reasons, it is worth studying Diophantus' solution, given here in the translation of Heath (1910).

8. To divide a given square number into two squares.

Given square number 16.
x^2 one of the required squares. Therefore $16 - x^2$ must be equal to a square.

Take a square of the form $(mx - 4)^2$, m being any integer and 4 the number which is the square root of 16, e.g. take $(2x - 4)$, and equate it to $16 - x^2$.

Therefore $4x^2 - 16x + 16 = 16 - x^2$,

or $5x^2 = 16x$, and $x = 16/5$.

The required squares are therefore $\frac{256}{25}$, $\frac{144}{25}$.

We see from this why Diophantus seeks *rational* solutions of equations. There are no positive integer squares x^2 and y^2 that sum to 16, so rational solutions are the interesting ones in this problem. But why try to express $16 - x^2$ as a square of the form $(mx - 4)^2$? It works algebraically because the constant term in $(mx - 4)^2$ cancels the 16, but $mx - 4$ also has a *geometric meaning*, which makes the solution easier to understand and generalize.

The pairs of numbers (x, y) such that $x^2 + y^2 = 16$ form a circle in the (x, y) plane, so Diophantus' problem is equivalent to finding *rational points* on this circle. Because $16 = 4^2$, there are some obvious rational points, for example, $x = 0$, $y = -4$. And $y = mx - 4$ *is a line through the "obvious" rational point* (0, 4). Diophantus is simply finding the *other* intersection of this line with the circle, in this case $m = 2$. He could choose any rational value of m and still find the other intersection to be rational.

By implication, Diophantus allows any rational value of m, so he can actually find *all* rational points on the circle, simply because the line through any rational point (s, t) and $(0, -4)$ has rational slope $m = \frac{t+4}{s}$. There is also nothing special about the radius 4. The rational points on a circle of any rational radius r can be found by multiplying those on the circle of radius 4 by $r/4$. Thus Diophantus has really solved the problem of finding all rational points on a circle of rational radius, as we claimed in Section 4.3.

He has also solved the equivalent problem: to divide a given (rational) square into two (rational) squares. It was this solution that inspired Fermat to make a note in the margin next to Problem 8 of Book II in his copy of Diophantus:

> It is impossible to separate a cube into two cubes, or a bi-quadrate into two biquadrates, or in general any power higher than second into powers of like degree: I have discovered a truly marvellous proof of this which however this margin is too small to contain.

More concisely, Fermat's claim is that the equation $x^n + y^n = z^n$ has no solution in positive integers x, y, z when n is an integer > 2. This became known as *Fermat's last theorem*, not because Fermat proved it, but because it was the last of Fermat's claims to be settled. In fact, Fermat was almost certainly mistaken to think he had a proof, though he could prove the case of biquadrates (fourth powers), as we saw in Section 4.7*.

Elliptic Curves

Fermat's last theorem was not proved until 1994, and then only through the work of several mathematicians: Gerhardt Frey, Jean-Pierre Serre, Ken Ribet, Richard Taylor, and especially Andrew Wiles. The proof involves some of the most abstract and difficult techniques of modern mathematics, but they are used to make a connection between the nth-degree equation $x^n + y^n = z^n$ and something relatively simple: a *cubic* equation of the form $y^2 = x(x-\alpha)(x-\beta)$. In 1984, Frey had the wild idea to suppose (contrary to Fermat's last theorem) that there are positive integers a, b, c with $a^n + b^n = c^n$, and to see what this implied about the curve with equation $y^2 = x(x - a^n)(x + c^n)$. He suspected, but could not prove, that the unlikely numbers a^n and c^n would give the curve an unlikely property, known as *nonmodularity*.

To cut a long story short, Fermat's last theorem was proved by showing that a counterexample (a, b, c) to Fermat's last theorem *does* imply nonmodularity (Serre and Ribet), but that nonmodularity is impossible for the curves $y^2 = x(x - \alpha)(x - \beta)$ (Taylor and Wiles). Consequently, there is no counterexample to Fermat's last theorem! It is way beyond the scope of this book to explain what *nonmodularity* is, but it is worth saying a few words about the cubic curves $y^2 = x(x - \alpha)(x - \beta)$, as they also go back to Diophantus and Fermat.

There is an important difference between quadratic and higher-degree curves, as we know from the exercises in Sections 4.3, 4.5, and 4.7*. Any quadratic curve can be parameterized by rational functions, but a higher degree curve generally can not. The simplest functions that can parameterize the cubic curve $y^2 = x(x - \alpha)(x - \beta)$ when $0 \neq \alpha \neq \beta$ are called *elliptic functions*, and for this reason we call $y^2 = x(x - \alpha)(x - \beta)$ an *elliptic curve*. (The elliptic curves also include some fourth-degree curves, such as $y^2 = 1 + x^4$. This curve can be parameterized by elliptic functions but, as we know from Exercise 4.7.5*, not by rational functions.)

Despite this, it is not that hard to find rational *points* on a cubic curve $\mathcal{K}$, provided the equation of $\mathcal{K}$ has rational coefficients. A simple argument, like the one given for quadratic curves in Section 4.5, shows that a line through two rational points on a cubic $\mathcal{K}$ with rational coefficients meets $\mathcal{K}$ in a third rational point. It is not even necessary to find two rational points to get started; one is enough, because the tangent at one rational point P effectively "meets $\mathcal{K}$ twice" at P, and hence its other intersection with $\mathcal{K}$ is also rational.

The algebraic equivalent of the tangent construction was actually used by Diophantus. In his Problem 18 of Book VI he uses the obvious solution $x = 0$, $y = 1$ of the equation $y^2 = x^3 - 3x^2 + 3x + 1$ to find the nonobvious solution $x = 21/4$, $y = 71/8$, by substituting $y = 3x/2 + 1$. The latter equation represents the tangent at $(0, 1)$.

Fermat took up Diophantus' tangent method to find rational solutions of cubic equations, and Newton pointed out the related method (the "chord construction") of drawing a line through two rational points to find a third. Finally, in 1922 Louis Mordell proved that these two methods suffice to find *all* rational points on a cubic curve, provided finitely many rational points are given. Mordell's theorem is difficult and deeply dependent on elliptic functions; nevertheless it shows that elliptic curves are near relatives of quadratic curves when it comes to finding rational points.

For this reason and because they are related to many classical problems, elliptic curves have been intensely studied over recent decades. The proof of Fermat's last theorem is the most spectacular result of this study so far, but others can be expected. The book of Koblitz (1985) is an attractive introduction to the subject, organized around an ancient problem that is still not solved: which integers are the areas of rational right-angled triangles?

5 Trigonometry

CHAPTER

5.1 Angle Measure

The word *trigonometry* comes from the Greek for "triangle measurement." More specifically, it means the study of relationships between the size of sides and the size of angles in triangles. Euclid says very little about this. He has theorems about equal angles and the sum of angles, and one angle being twice another or simply larger than another, but he never actually *measures* angles. He does not represent angles by numbers, nor does he represent them by lengths or areas. This suggests that angle measure may be a deep concept, perhaps beyond the scope of traditional geometry. The Greeks had some inkling of this when they tried unsuccessfully to construct the area bounded by the unit circle, the problem they called *squaring the circle*. In modern terms, squaring the circle amounts to constructing the number π, which is both the area of the unit circle and half its circumference. It is also the natural measure of the straight angle, formed by two right angles, so constructing π is in fact a fundamental question about the measurement of angles.

Probably the only way to understand π well enough to know whether it is constructible is to use advanced calculus, which is beyond the scope of this book. However, we can understand angle

143

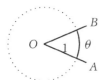

FIGURE 5.1   Representing an angle by an arc.

measure with less; the concept can be made clear with the help
of analytic geometry and the theory of real numbers developed in
Chapter 3. We shall also find that this is enough to capture the elusive
number π in the form of an infinite sum or product.

Let us begin with Euclid's idea of angle, as a pair of rays OA
and OB we call angle AOB. The implication of this notation is that
OA is the first ray in the angle and OB is the second. But there is
still the problem of explaining which way to travel from OA to OB—
clockwise or counterclockwise (recall the discussion of orientation
in Chapter 3). The easiest way out of this problem and several others,
is to draw a unit circle centered on O and to choose one of the arcs
between the rays to mark the intended angle (Figure 5.1, with the
chosen arc drawn heavily).

The *measure* of the angle AOB can then be defined as the *length*
θ of the arc AB. We have not yet defined length of arcs, admittedly,
but this is not hard. The length of any arc between points P and
Q on the circle may be defined as the least upper bound of the
length of polygons $P_1P_2\ldots P_{n-1}P_n$ joining points $P_1, P_2, \ldots P_{n-1}, P_n$
that lie in that order on the arc between P and Q (Figure 5.2). As
implied in the exercises in Section 3.7, this least upper bound exists
because a polygon joining points on the circle is shorter than a square
enclosing the circle, so there is an upper bound to the set of polygonal

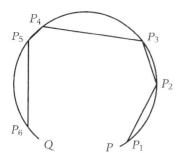

FIGURE 5.2   Arc and polygon.

lengths and hence a least upper bound by the completeness of the real numbers.

Finally, we define the number π to be the length of a semicircle of radius 1. It will be some time before we can give a precise value of π, but in the meantime we need to know what we are talking about when we say that the length of the whole unit circle is 2π and the like.

Exercises

In the early history of π, some very rough estimates were used. For example, there is a verse in the Bible (Kings 7:23) about a "molten sea, ten cubits from the one brim to the other: it was round all about ... and a line of thirty cubits did compass it round about." If the sea was circular, this assumes $\pi = 3$. This value is easily seen to be too small.

5.1.1. By inscribing a regular hexagon in a circle, show that $\pi > 3$.

The idea of approximating the circle by polygons dominated the study of π from the time of Archimedes (around 250 B.C.) until about 1500 A.D. Polygons inside and outside the circle were used to narrow the interval in which π was known to lie.

5.1.2. Use the squares in Figure 5.3 to show that $2\sqrt{2} < \pi < 4$ and the octagons in Figure 5.3 to show that $4\sqrt{2 - \sqrt{2}} < \pi < 8(\sqrt{2} - 1)$. (The latter approximations give $3.06 < \pi < 3.32$.)

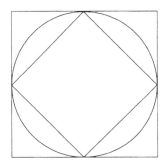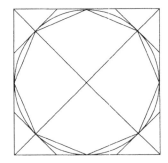

FIGURE 5.3   Approximating the circle by squares and octagons.

The first accurate bounds on π were found by Archimedes, who used inner and outer polygons with 96 sides to show that

$$3\frac{10}{71} < \pi < 3\frac{1}{7}.$$

This neat improvement on the school value of $3\frac{1}{7}$ gives the decimal estimates $3.140 < \pi < 3.143$, and hence gives π correct to two decimal places.

Outer polygons are not needed to *define* the length of the circle, except to ensure that there is an upper bound to the length of polygons inside the circle, because the least upper bound of the lengths of inner polygons exists by the completeness of the real numbers. However, they are useful for finding how close a given inner polygon comes to the circle; it is closer to the circle than to any outer polygon.

5.1.3. Use the triangle inequality to show that *any* polygon inside the circle is shorter than every polygon outside the circle.

Outer polygons also assure us that it is sensible to define the length of the circle as the least upper bound of lengths of inner polygons, because we can show that the difference in length between inner and outer polygons can be made as small as we please. This means it is equivalent to use the (equally natural) definition that the length of the circle is the greatest lower bound of the lengths of outer polygons.

Figure 5.4 helps to explain why the difference in length can be made as small as we please. It shows a sector of the unit circle, the half-side y of an inner polygon, and the half-side $y + \varepsilon$ of an outer polygon.

5.1.4. Show that $\frac{y}{y+\varepsilon} = 1 - \delta$. Hence conclude that, by suitable choice of δ, we can make the ratio of lengths of an inner and outer polygon as close to 1 as we please. This implies that the difference between their lengths can be made as close to 0 as we please. Why?

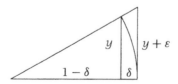

FIGURE 5.4   Sides of inner and outer polygons.

These inner and outer approximations are analogous to upper and lower sets of a Dedekind cut. And like them, they make it easy to define the *sum* of angles. Recall that our purpose in defining arc length on the unit circle was to define angle measure. We now want to see whether the measure of a sum of angles is the sum of their measures. When angles are added by joining them along a common ray, the corresponding arcs are joined at a common point. One certainly expects the length of the combined arc to be the sum of the lengths of the two pieces, and in fact this follows from the definition of arc length as a least upper bound.

5.1.5. Show the following, for arcs *AB*, *BC*, and their combined arc *AC*:

- Any polygons drawn inside *AB* and *BC* have total length < some polygon drawn inside *AC*.

- Any polygon drawn inside *AC* has length < the sum of the lengths of some polygons drawn inside *AB* and *BC*.

- It follows that length(*AC*) = length(*AB*) + length(*BC*).

5.2   Circular Functions

We first meet the circular functions sine and cosine at school, as ratios of sides of right-angled triangles (Figure 5.5). These ratios depend only on the size θ of the angle, and not the size of the triangle, because of the basic property of straight lines, in fact, by the *defining* property of straight lines in analytic geometry (Section 3.1):

$$\cos\theta = \frac{a}{c}, \qquad \sin\theta = \frac{b}{c}.$$

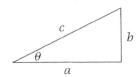

FIGURE 5.5   Defining cos and sin via a triangle.

In this context, cos and sin are called *trigonometric* functions, because they assist in the measurement of triangles. However, this definition limits their domain to angles θ less than $\pi/2$, which is inconvenient for at least two reasons:

- There are formulas for $\cos(\theta+\phi)$ and $\sin(\theta+\phi)$ in terms of $\cos\theta$, $\cos\phi$, $\sin\theta$, $\sin\phi$, which suggest a meaning for $\cos(\theta + \phi)$ and $\sin(\theta + \phi)$ when $\theta + \phi > \pi/2$.

- The functions cos and sin not only give sides of triangles as functions of angle, but also amplitude of vibration as a function of time or the height of a wave as a function of distance—and the time or distance can be *any* real number.

This leads us to extend the definition of cos and sin so that they make sense for any real number θ. It then becomes more appropriate to call them *circular* functions.

We take the unit circle (Figure 5.6) and view the coordinates x and y of any point P on it as functions $\cos\theta$ and $\sin\theta$ of the angle θ. This gives a meaning to $\cos\theta$ and $\sin\theta$ for all θ from 0 to 2π, but there is no reason to stop there. It is natural to define cos and sin for θ outside this interval by the equations

$$\cos(\theta + 2\pi) = \cos\theta,$$
$$\sin(\theta + 2\pi) = \sin\theta,$$

because increasing θ by 2π means making a complete circuit, and hence returning to the same point P on the circle. Likewise, there is no reason to distinguish between "angle θ" and "angle $\theta + 2\pi$." Any two real numbers that differ by 2π represent the same angle, so an

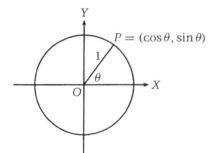

FIGURE 5.6   Defining cos and sin via the circle.

angle is really a *set* of real numbers of the form

$$\{\theta + 2n\pi : n \in \mathbb{Z}\} = \{\dots, \theta - 4\pi, \theta - 2\pi, \theta, \theta + 2\pi, \theta + 4\pi, \dots\},$$

obtained by adding all integer multiples of 2π to θ. As in previous cases (such as Dedekind cuts), there are advantages to defining a mathematical object as a set. We now have no problem defining the sum of angles:[1] the *sum* of the angle $\{\theta + 2n\pi : n \in \mathbb{Z}\}$ and the angle $\{\phi + 2n\pi : n \in \mathbb{Z}\}$ is simply the angle $\{\theta + \phi + 2n\pi : n \in \mathbb{Z}\}$.

When cos and sin are related to the circle in this way, it becomes obvious why they are relevant to rotation and vibration. If the point P travels around the circle at constant angular velocity, so that θ measures time as well as angle, then $x = \cos\theta$ and $y = \sin\theta$ measure the horizontal and vertical displacements of the uniformly rotating point. The x-coordinate of P can be viewed as the position of its shadow under a light shining vertically downward, and movement of this shadow is the simplest form of vibration. It is called *simple harmonic motion* because such vibration is the basis of musical tones.

It is also obvious, from the circular interpretation of cos and sin, that their graphs have the same shape and that the graph of sin lags $\pi/2$ behind the graph of cos. (Figure 5.7 shows $y = \sin x$ drawn heavily and $y = \cos x$ drawn dotted). The shape is known as the *sine wave*.

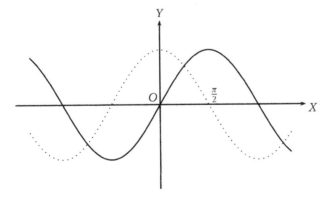

FIGURE 5.7
Graphs of the cos and sin functions.

---

[1] Note, however, that when we speak of the "angle sum" of a triangle, quadrilateral, and so on, we take the angles to be real numbers between 0 and 2π, and we use the sum of reals.

Exercises

The properties of cos and sin bear out the claim that the concept of angle measure lies outside elementary geometry. In fact, cos and sin are *transcendental functions*, which lie outside the realm of *algebraic functions* we have considered so far. In general, a function $y(x)$ is called algebraic if $p(x, y) = 0$ for some polynomial p in the two variables x and y. The graph $p(x, y) = 0$ of an algebraic function is called an *algebraic curve*.

For example, $y = \sqrt{1 - x^2}$ is an algebraic function of x, because it satisfies the equation

$$x^2 + y^2 = 1,$$

which is of the form $p(x, y) = 0$, with $p(x, y) = x^2 + y^2 - 1$. In this case the algebraic curve is simply the unit circle.

The curve $y = \sin x$ is *not* an algebraic curve, and hence $\sin x$ is not an algebraic function. The reason is that the sine curve meets the line $y = 0$ infinitely often, namely, at the points $x = n\pi$ for all integers n. An algebraic curve $p(x, y) = 0$, on the other hand, meets the line $y = mx + c$ where $p(x, mx + c) = 0$, which is a polynomial equation whose roots are the x-coordinates of the points of intersection. Such an equation cannot have infinitely many roots. In fact, a polynomial equation of degree n can have at most n roots. The following exercises give one way to see this.

5.2.1. Check that $x^n - a^n = (x - a)(x^{n-1} + x^{n-2}a + \cdots + ax^{n-2} + x^{n-1})$.

5.2.2. Deduce from Exercise 5.2.1 that if $p(x)$ is polynomial of degree n, then

$$p(x) - p(a) = (x - a)q(x),$$

where $q(x)$ is polynomial of degree $n - 1$.

5.2.3. Deduce from Exercise 5.2.2 that if $p(x)$ is a polynomial of degree $n > 0$, then $p(a)$ cannot be zero for more than n different numbers a.

Thus cos and sin are examples of transcendental functions. The same is true of any function f that satisfies an equation of the form

$$f(x + \alpha) = f(x) \quad \text{for some } \alpha \neq 0,$$

because its graph meets the horizontal line $y = f(\alpha)$ for infinitely many values of x. We call such a function *periodic*, with *period* α. Many other periodic functions can be built from cos and sin, for example,

$$\tan x = \frac{\sin x}{\cos x} \quad \text{and} \quad \cot x = \frac{\cos x}{\sin x}.$$

The cotangent function, cot, is noteworthy because Euler (1748) discovered a formula that "shows" its period, namely,

$$\pi \cot \pi x = \cdots + \frac{1}{x-2} + \frac{1}{x-1} + \frac{1}{x} + \frac{1}{x+1} + \frac{1}{x+2} + \cdots.$$

5.2.4. Give a geometric reason why $\cot \pi x$ has period 1. Why does Euler's formula show that $\pi \cot \pi (x + 1) = \pi \cot \pi x$?

5.2.5. Euler's formula suggests that $\cot \pi x$ tends to infinity as x approaches any integer value. Give a geometric explanation of this behavior.

Euler's formula for $\pi \cot \pi x$ is probably the simplest one can imagine that shows periodicity, so there may be a sense in which the cot function is more fundamental than cos and sin. In fact, we shall see in Section 5.3 that either tan or cot may be used as a "primitive" circular function, with both cos and sin defined in terms of it.

Formulas for circular functions are most easily derived using calculus, but it takes time to build calculus to the point where it works efficiently. Instead, we shall get by with a few *limit properties* of the circular functions, the most important of which is:

$$\lim_{\theta \to 0} \frac{\sin \theta}{\theta} = 1.$$

This means we can ensure that $\sin \theta / \theta$ is within any given positive distance (say, ε) of 1, by choosing θ sufficiently small (say, $< \delta$).

5.2.6. By referring to Figure 5.8 and the definition of angle measure, show that

$$\sin \theta < \theta < \tan \theta \quad \text{and hence that} \quad \cos \theta < \frac{\sin \theta}{\theta} < 1.$$

5.2.7. Deduce from Exercise 5.2.6 that $\lim_{\theta \to 0} (\sin \theta)/\theta = 1$.

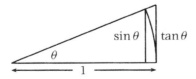

FIGURE 5.8 Comparing sin, arc, and tan.

5.3 Addition Formulas

The functions cos and sin are necessarily complicated, inasmuch as they are transcendental, but there are still some simple relations between them. For example, because $x = \cos\theta$ and $y = \sin\theta$ are the coordinates of a point (x, y) on the circle $x^2 + y^2 = 1$, we necessarily have

$$(\cos\theta)^2 + (\sin\theta)^2 = 1.$$

We usually write this

$$\cos^2\theta + \sin^2\theta = 1,$$

even though the notation $\cos^2\theta$ conflicts with the notation $\cos^{-1}\theta$ for the inverse cosine. When there is a danger of misunderstanding, it is wise to write $(\cos\theta)^2$ for the square of $\cos\theta$.

This relation enables us to express either one of $\cos\theta$ or $\sin\theta$ in terms of the other, namely,

$$\cos\theta = \sqrt{1 - \sin^2\theta} \quad \text{and} \quad \sin\theta = \sqrt{1 - \cos^2\theta}.$$

There is a cost, however, because now we have to worry about the sign of the square root. As we shall see in the next section, there are unambiguous formulas for both $\cos\theta$ and $\sin\theta$ in terms of $\tan\frac{\theta}{2}$.

After $\cos^2\theta + \sin^2\theta = 1$, the most important relations between cos and sin are the so-called *addition formulas*:

$$\cos(\theta + \phi) = \cos\theta\cos\phi - \sin\theta\sin\phi,$$
$$\sin(\theta + \phi) = \sin\theta\cos\phi + \cos\theta\sin\phi.$$

We prove just the first of these, because the second is similar (and in fact it follows from the first). The proof refers to Figure 5.9.

Looking first at the right-angled triangle OAC, we see

$$OA = \cos\phi \quad \text{and} \quad AC = \sin\phi.$$

Next, viewing OA and AC as the hypotenuses of the right-angled triangles ODA and ABC, respectively, we see

$$OD = \cos\theta\cos\phi \quad \text{and} \quad AB = \sin\theta\sin\phi$$

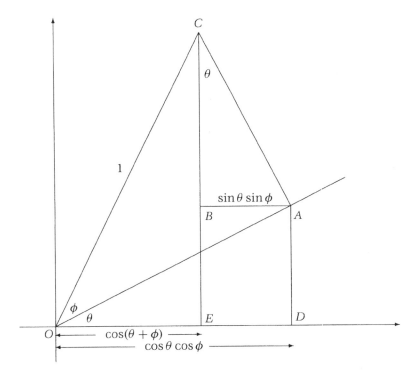

FIGURE 5.9 Constructing the cosine of a sum.

(the latter because angle $ACB = \theta$, because angle $OAC = \pi/2$). This finally gives

$$\cos(\theta + \phi) = OE = OD - AB = \cos\theta\cos\phi - \sin\theta\sin\phi,$$

which is the required result. $\qquad\qquad\square$

Exercises

The addition formula for cosine is useful, but not quite as simple or memorable as one would like. The same goes for the addition formula for sine.

5.3.1. Prove the addition formula for sine, $\sin(\theta + \phi) = \sin\theta\cos\phi + \cos\theta\sin\phi$,

- by using Figure 5.9, but with appropriate vertical lengths instead of horizontal lengths, or

- by deducing it from the addition formula for cosine, with the help of the formulas $\sin\alpha = \cos(\frac{\pi}{2} - \alpha)$ and $\cos\alpha = \sin(\frac{\pi}{2} - \alpha)$.

By a kind of miracle, these two somewhat complicated formulas are parts of one simple formula. To state it, we have to use the "imaginary number" $i = \sqrt{-1}$, which will be explained more fully in Chapter 7. For the moment it is enough to know that $i^2 = -1$ and that if $\alpha + i\beta = \gamma + i\delta$ for real numbers $\alpha, \beta, \gamma, \delta$ then $\alpha = \gamma$ and $\beta = \delta$. Then the famous *de Moivre formula* (1730) is:

$$\cos(\theta + \phi) + i\sin(\theta + \phi) = (\cos\theta + i\sin\theta)(\cos\phi + i\sin\phi).$$

5.3.2. Verify the de Moivre formula, by multiplying out the right-hand side and using the addition formulas for cos and sin.

The de Moivre formula is so simple, compared with the cosine and sine formulas, that it seems that the function $\cos\theta + i\sin\theta$ is simpler than either its "real part" $\cos\theta$ or its "imaginary part" $\sin\theta$. Indeed, if we abbreviate $\cos\theta + i\sin\theta$ by $\operatorname{cis}\theta$, the addition formulas for cos and sin unite in the spectacularly simple addition formula for cis:

$$\operatorname{cis}(\theta + \phi) = \operatorname{cis}\theta \cdot \operatorname{cis}\theta.$$

If this reminds you of the exponential function, it should! However, we are getting ahead of the story. First, there are some important applications of the addition formulas to consider.

An easy special case is $\cos 2\theta = \cos^2\theta - \sin^2\theta$, which can be rewritten using $\cos^2\theta + \sin^2\theta = 1$ as $\cos 2\theta = 2\cos^2\theta - 1 = 1 - 2\sin^2\theta$.

5.3.3. From $\cos 2\theta = 1 - 2\sin^2\theta$ deduce $1 - \cos\theta = \sin^2\frac{\theta}{2}$, and hence show that $\lim_{\theta\to 0}(1 - \cos\theta)/\theta = 0$ with the help of the result $\lim_{\theta\to 0}(\sin\theta)/\theta = 1$ from Exercise 5.2.7.

By combining these limit results with the sine addition formula, we can find the tangent to the sine wave at any point, which in calculus is called "finding the derivative of $\sin x$." The tangent to any curve at a point P, if it exists, is found as the limiting position of a chord between P and a point $Q \neq P$, as Q approaches P. In the case of the sine wave $y = \sin x$, we take $P = (\alpha, \sin\alpha)$, $Q = (\alpha + \theta, \sin(\alpha + \theta))$, and let $\theta \to 0$.

5.3.4. Show that the slope of the chord between P and Q is

$$\frac{\sin(\alpha + \theta) - \sin\alpha}{\theta} = \frac{\sin\alpha(\cos\theta - 1) + \cos\alpha\sin\theta}{\theta},$$

and deduce that the slope of the tangent at $x = \alpha$ is $\cos\alpha$.

5.3.5. Similarly use the cosine addition formula to show that the slope of the tangent to $y = \cos x$ at $x = \alpha$ is $-\sin\alpha$.

The de Moivre formula also makes it easy to find formulas for $\cos n\theta$ and $\sin n\theta$. It is as easy as expanding $(\cos\theta + i\sin\theta)^n$.

5.3.6. By expanding $(\cos\theta + i\sin\theta)^3$, show that

$$\cos 3\theta = 4\cos^3\theta - 3\cos\theta,$$

$$\sin 3\theta = 3\sin\theta - 4\sin^3\theta.$$

5.3.7. Show that $\cos n\theta$ is a polynomial in $\cos\theta$, for any natural number n. What is the situation for $\sin n\theta$?

These polynomials were discovered by Viète (1579) and were used by him to solve certain polynomial equations by circular functions. In 1593 he won a mathematical contest by noticing that a 45th-degree equation posed by his opponent was based on the polynomial for $\sin 45\theta$.

Another famous discovery of Viète is also based on addition formulas: his infinite product

$$\frac{2}{\pi} = \cos\frac{\pi}{4}\cos\frac{\pi}{8}\cos\frac{\pi}{16}\cdots$$

$$= \frac{\sqrt{2}}{2}\frac{\sqrt{2+\sqrt{2}}}{2}\frac{\sqrt{2+\sqrt{2+\sqrt{2}}}}{2}\cdots.$$

5.3.8.* Use the sine addition formula to show in turn that

$$\sin\theta = 2\sin\frac{\theta}{2}\cos\frac{\theta}{2},$$

$$\frac{\sin\theta}{2^n\sin(\theta/2^n)} = \cos\frac{\theta}{2}\cos\frac{\theta}{2^2}\cdots\cos\frac{\theta}{2^n},$$

$$\frac{\sin\theta}{\theta} = \cos\frac{\theta}{2}\cos\frac{\theta}{2^2}\cos\frac{\theta}{2^3}\cdots,$$

and deduce Viète's product by substituting $\theta = \pi/2$.

5.4 A Rational Addition Formula

The addition formulas for cos and sin give the *double angle* formulas

$$\cos 2\theta = \cos^2 \theta - \sin^2 \theta,$$
$$\sin 2\theta = 2 \sin \theta \cos \theta.$$

By rewriting $\cos 2\theta$ we find

$$\cos 2\theta = \frac{\cos^2 \theta - \sin^2 \theta}{\cos^2 \theta + \sin^2 \theta} \quad \text{because } \cos^2 \theta + \sin^2 \theta = 1$$
$$= \frac{1 - \tan^2 \theta}{1 + \tan^2 \theta}, \quad \text{dividing numerator and denominator by } \cos^2 \theta.$$

Similarly,

$$\sin 2\theta = \frac{2 \sin \theta \cos \theta}{\cos^2 \theta + \sin^2 \theta} \quad \text{because } \cos^2 \theta + \sin^2 \theta = 1$$
$$= \frac{2 \tan \theta}{1 + \tan^2 \theta}, \quad \text{dividing numerator and denominator by } \cos^2 \theta.$$

Finally, replacing θ by $\theta/2$, we get the *half angle formulas* expressing $\cos \theta$ and $\sin \theta$ rationally in terms of $\tan \frac{\theta}{2}$:

$$\cos \theta = \frac{1 - \tan^2 \frac{\theta}{2}}{1 + \tan^2 \frac{\theta}{2}}, \qquad \sin \theta = \frac{2 \tan \frac{\theta}{2}}{1 + \tan^2 \frac{\theta}{2}}.$$

This supports our claim from Section 5.2 that the tan function may be considered more fundamental than either cos or sin. The surprise is that we already know these formulas! They are essentially the formulas used by Diophantus to find rational Pythagorean triples. Look again at the diagram we used in Section 4.2 to explain Diophantus' construction, and the role of $\tan \frac{\theta}{2}$ becomes clear (Figure 5.10).

If we take the angle to the point $(\frac{1-t^2}{1+t^2}, \frac{2t}{1+t^2})$ to be θ as shown, then the line from $(-1, 0)$ to the same point is at angle $\frac{\theta}{2}$, by the theorem that the angle at the circumference is half the angle at the center (Section 2.4). It follows that the slope t of the line is $\tan \frac{\theta}{2}$, and the coordinates $\frac{1-t^2}{1+t^2}$ and $\frac{2t}{1+t^2}$ are $\cos \theta$ and $\sin \theta$, respectively.

This prompts the thought that we should be able to add angles by calculating the corresponding slopes, and hence work with rational functions instead of the transcendental functions cos and sin. What

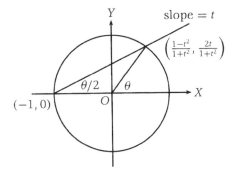

FIGURE 5.10 The half angle construction.

we need is the *addition formula for tan,* which is found as follows:

$$
\begin{aligned}
\tan(\theta + \phi) &= \frac{\sin(\theta + \phi)}{\cos(\theta + \phi)} \\
&= \frac{\sin\theta\cos\phi + \cos\theta\sin\phi}{\cos\theta\cos\phi - \sin\theta\sin\phi} \\
&= \frac{\tan\theta + \tan\phi}{1 - \tan\theta\tan\phi},
\end{aligned}
$$

dividing numerator and denominator by $\cos\theta\cos\phi$. This formula gives us what we want:

Rational addition formula. *If the line through O at angle θ has slope s and the line at angle ϕ has slope t, then the line at angle $\theta + \phi$ has slope $(s + t)/(1 - st)$.* □

With this formula, we can finally solve the problem the Babylonians were trying to solve with their table of Pythagorean triples (Section 4.1). In effect, they were looking for equally spaced rational points on an arc of the unit circle. Any number of them can now be found, by starting with a point that has a small angle θ and a rational slope relative to O, and finding the points at angles 2θ, 3θ, and so on. By the rational addition formula, these points also have rational slopes relative to O, and hence they are rational points.

For example, suppose we start with the Pythagorean triple $(24, 7, 25)$. This corresponds to the rational point at slope $s = 7/24$ and angle θ of about $16°$. By the addition formula for tan, the point at angle 2θ has slope

$$t = \frac{2s}{1 - s^2} = \frac{2 \times 7/24}{1 - 7^2/24^2} = \frac{14 \times 24}{24^2 - 7^2} = \frac{336}{527},$$

and this point corresponds to the Pythagorean triple $(527, 336, 625)$. The point at angle 3θ has slope

$$u = \frac{s+t}{1 - st} = \frac{\frac{7}{24} + \frac{336}{527}}{1 - \frac{7}{24}\frac{336}{527}} = \frac{7 \times 527 + 336 \times 24}{24 \times 527 - 7 \times 336} = \frac{11753}{10296},$$

and this corresponds to the Pythagorean triple $(10296, 11753, 15625)$. Clearly the process can be continued indefinitely, or at least until the computations become unmanageable.

We never return to the initial rational point on the circle by continuing this process. Hence, it is impossible to improve our solution of the Babylonian problem by subdividing a right angle, say, with equally spaced rational points. In fact, it is impossible to divide the circle into more than four equal parts by rational points. The exercises to Section 5.8* will explain why.

Exercises

Having seen how $i = \sqrt{-1}$ simplifies the addition formulas for cos and sin, we would also expect it to help with Pythagorean triples. This expectation is fulfilled. When one reflects on the connections between triples, points on the unit circle, and angles, the following procedure comes naturally to mind:

- Replace the triple (a, b, c) by the point $(a/c, b/c)$ on the unit circle.

- Think of the point $(a/c, b/c)$ as $(\cos\theta, \sin\theta)$, which in turn can be replaced by the number $\cos\theta + i\sin\theta$.

- Given two triples (a_1, b_1, c_1) and (a_2, b_2, c_2), form the corresponding numbers $a_1/c_1 + ib_1/c_1$ and $a_2/c_2 + ib_2/c_2$. The product $a_3/c_3 + ib_3/c_3$ of the latter numbers yields a new Pythagorean triple (a_3, b_3, c_3).

5.4.1. Show that if (a_1, b_1, c_1) and (a_2, b_2, c_2) are Pythagorean triples, then so is their "product" (a_3, b_3, c_3), where a_3 and b_3 are defined by

$$a_3 + ib_3 = (a_1 + ib_1)(a_2 + ib_2)$$

and $c_3 = \sqrt{a_3^2 + b_3^2}$.

5.4.2. Show that $c_3 = c_1 c_2$.

5.4.3. Show that the $(3, 4, 5)$ triple has "square" $(-7, 24, 25)$.

Conversely, one may "factorize" certain triples into products of simpler triples. Some of the triples (a, b, c) in Plimpton 322 "factorize" in this sense, though only a minority of them, because c is a prime number for most. A "factorization" of (a, b, c) implies that the angle with the rational slope b/a is a nontrivial sum of two angles with rational slope. In particular, if (a, b, c) is a "perfect square" then half its angle also has rational slope.

5.4.4. Show that the triples $(119, 120, 169)$ and $(161, 240, 289)$ from Plimpton 322 are "perfect squares" in this sense.

5.4.5. Can you suggest a method of "division" to find a second "factor" of a Pythagorean triple when one "factor" is known?

5.4.6. "Factorize" the following triples from Plimpton 322:

$$(319, 360, 481), \quad (1679, 2400, 2929), \quad (4601, 4800, 6649).$$

5.5* Hilbert's Third Problem

The rest of this chapter is concerned with a famous problem we met in Section 2.7: is it possible to cut a regular tetrahedron into finitely many pieces by planes and paste the pieces into a cube? As mentioned before, this problem is the main obstacle when we attempt to develop a theory of volume using only finite processes. Its importance was recognized by Hilbert, and he placed it at number 3 on the list of problems for 20th century mathematicians he announced in Paris in 1900. It was solved a few months later by Hilbert's student Max Dehn, with surprisingly simple methods. It is certainly the only one of Hilbert's problems whose solution can be described in a book such as this, and it happens to be relevant here, because trigonometry plays an important role in it.

Dehn's solution comes from focusing on the *dihedral angles* of a polyhedron, the angles between its faces. If a tetrahedron can be cut up and pasted into a cube, for example, then it looks like we have to build right angles (the dihedral angles of a cube) from the dihedral angles of a tetrahedron. It is not quite that simple, because the cuts can create dihedral angles in the interior, but one feels that these "cancel out" in some sense. Later we'll describe precisely how one keeps track of dihedral angles as a polyhedron is cut and pasted, but the first step is to actually find the dihedral angles of a tetrahedron. This is where trigonometry makes its first appearance in the problem; we have to measure the triangle ABC in which angle ABC is the dihedral angle of the tetrahedron (Figure 5.11).

It follows from Pythagoras' theorem that $AB = BC = \sqrt{3}/2$, and this in turn implies $BD = 1/\sqrt{2}$. Consequently, if α is the dihedral angle,

$$\cos\frac{\alpha}{2} = \frac{BD}{AB} = \frac{2}{\sqrt{2}\sqrt{3}},$$

and therefore

$$\cos\alpha = 2\cos^2\frac{\alpha}{2} - 1 = \frac{2\times 4}{6} - 1 = \frac{1}{3}.$$

Thus one of the things we have to understand is the relationship between $\pi/2$, the dihedral angle of the cube, and the angle whose cosine is $1/3$. The keys to this relationship turn out to be the addition formula and some basic number theory, as we shall see in Section

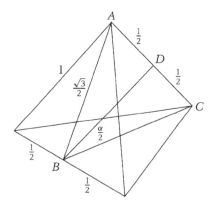

FIGURE 5.11   The dihedral angle of the regular tetrahedron.

5.8*. But first we need a better understanding of the behavior of a polyhedron under cutting and pasting, so that we can keep track of its dihedral angles.

Exercises

One of the nice properties of the angles of a polygon is that their sum is an integer multiple of π, in fact $(n-2)\pi$, where n is the number of vertices (Exercise 2.3.3). The dihedral angles of a polyhedron do not behave so nicely. Their sum is an integer multiple of π for some polyhedra, such as the cube, but for others it is not.

5.5.1. If α is the dihedral angle of the regular tetrahedron, show that $\cos 2\alpha = -7/9$ and $\cos 6\alpha = 329/729$.

5.5.2. Deduce from Exercise 5.5.1 that the dihedral angle sum of a regular tetrahedron is not an integer multiple of π.

5.6* The Dehn Invariant

Dehn solved Hilbert's third problem by a stroke of genius. He saw that volume is not the only thing conserved by cutting and pasting a polyhedron. Another is what might be called its *dihedral content*, an object that encodes the dihedral angles and ties them to the lengths of the corresponding edges.

An edge of length l and dihedral angle α makes a contribution to the dihedral content that is written $l \otimes \alpha$. The total dihedral content of a polyhedron (or a finite set S of polyhedra) is written

$$D(S) = l_1 \otimes \alpha_1 + l_2 \otimes \alpha_2 + \cdots + l_k \otimes \alpha_k,$$

where $l_1, l_2, \ldots, l_k$ are the lengths of the edges and $\alpha_1, \alpha_2, \ldots, \alpha_k$ are their respective dihedral angles. Because the grouping or order of the length $\otimes$ angle pairs does not matter, $+$ is an associative and commutative operation, and there is no harm in confusing it with ordinary addition.

However, $D(S)$ is so far just an expression containing some information about a *single* finite set S of polyhedra. If $D(S)$ is also to describe the various sets S', S'', ... obtainable from S by cutting and pasting, we shall need rules that transform $D(S)$ into $D(S')$, $D(S'')$, ... These rules are very easy to state:

$$l \otimes (\alpha + \beta) = l \otimes \alpha + l \otimes \beta \qquad \text{(Rule 1)}$$
$$(l + m) \otimes \alpha = l \otimes \alpha + m \otimes \alpha \qquad \text{(Rule 2)}$$
$$l \otimes \pi = 0 \qquad \text{(Rule 3)}$$

Rule 1 tells what to do when a cut is made along an edge, splitting its dihedral angle $\alpha + \beta$ into dihedral angles α and β (Figure 5.12). Conversely, it tells what to do when two dihedral angles are pasted into one.

Rule 2 tells what to do when a cut is made across an edge of length $l + m$, splitting it into edges of lengths l and m (Figure 5.13). Conversely, it tells what to do when two edges are joined end-to-end into one.

Rule 3 tells us that an edge with angle π can be ignored, as it should be, because it is not an actual edge. (One such spurious edge is produced, as in Figure 5.14, when we cut along an edge with dihedral angle $\pi + \alpha$, splitting off the dihedral angle α. This creates

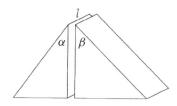

FIGURE 5.12 Why $l \otimes (\alpha + \beta) = l \otimes \alpha + l \otimes \beta$.

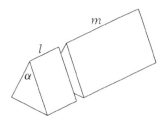

FIGURE 5.13 Why $(l + m) \otimes \alpha = l \otimes \alpha + m \otimes \alpha$.

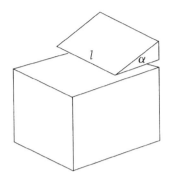

FIGURE 5.14   Why $l \otimes \pi = 0$.

an actual edge in one piece but not in the other, so $l \otimes (\pi + \alpha)$ is replaced by $l \otimes \alpha$.)

These figures are a little too simple, because they show the edge l perpendicular to the faces at its ends, so the dihedral angle is actually the angle visible at the end of l. But even if the dihedral angle is not visible, rules 1, 2, and 3 correctly express what happens to it under cutting and pasting.

When $D(S)$ is subjected to these rules it is called the *Dehn invariant*, because by definition it remains the same when S is cut or pasted. In particular, if P and Q are equidecomposable polyhedra, then $D(P) = D(Q)$.

Example   The Dehn invariant of the unit cube is 0.

This Dehn invariant is $12 \otimes \frac{\pi}{2}$, because the cube has 12 edges, each of length 1 and of dihedral angle $\pi/2$. But it follows from Rule 1 that

$$1 \otimes \frac{\pi}{2} + 1 \otimes \frac{\pi}{2} = 1 \otimes \left(\frac{\pi}{2} + \frac{\pi}{2} \right) = 1 \otimes \pi,$$

which equals 0 because $1 \otimes \pi = 0$ by Rule 3.   □

Rules 1 and 2 are so simple that systems obeying them have a name—*tensor products*—and have been studied in modern algebra. A consequence of Rule 1, for example, is that $l \otimes 0 = 0$, because

$$l \otimes \alpha = l \otimes (\alpha + 0) = l \otimes \alpha + l \otimes 0.$$

Because of this, Rule 3 should be regarded not as a property of the $\otimes$ operation but as a property of the set of dihedral angles. This set is denoted by $\mathbb{R}/\pi\mathbb{Z}$ and, informally speaking, it is what $\mathbb{R}$ becomes when we pretend that $\pi = 0$. Its members are actually the sets $\{\ldots, \alpha - 2\pi, \alpha - \pi, \alpha, \alpha + \pi, \alpha + 2\pi, \ldots\}$ for each real number α. $\mathbb{R}/\pi\mathbb{Z}$ is very like the set of angles, which in fact is $\mathbb{R}/2\pi\mathbb{Z}$. For angles, we always want $2\pi = 0$, but with the Dehn invariant we also want $\pi = 0$, because an edge with dihedral angle π is not an edge at all.

The objects $l_1 \otimes \alpha_1 + l_2 \otimes \alpha_2 + \cdots + l_k \otimes \alpha_k$ that occur as values of the Dehn invariant are today called *tensors*. The set of them is denoted by $\mathbb{R} \otimes \mathbb{R}/\pi\mathbb{Z}$ and is called the *tensor product of* $\mathbb{R}$ *and* $\mathbb{R}/\pi\mathbb{Z}$. Tensor products are normally studied in advanced algebra courses, where more sophisticated methods are available. However, we shall be able to prove what we need about the Dehn invariant from first principles, as Dehn himself did.

Exercises

The simpler tensor product $\mathbb{R} \otimes \mathbb{R}$ is also related to a decomposition problem. Consider a set of rectangles with horizontal and vertical sides, and suppose that the rectangles can be cut and pasted along vertical and horizontal lines. We represent a single rectangle with horizontal side x and vertical side y by the tensor $x \otimes y$ and a set of them by a sum of such terms. If a vertical cut divides the $(x_1 + x_2) \otimes y$ rectangle into two, of width x_1 and x_2, respectively, then we have the rule

$$(x_1 + x_2) \otimes y = x_1 \otimes y + x_2 \otimes y.$$

Similarly, a horizontal cut yields the rule

$$x \otimes (y_1 + y_2) = x \otimes y_1 + x \otimes y_2.$$

These rules define the tensor product $\mathbb{R} \otimes \mathbb{R}$. Thus each member of $\mathbb{R} \otimes \mathbb{R}$ may be interpreted as the set of all sums of rectangles that are equivalent under vertical and horizontal cut and paste. In particular, if $x \otimes y = x' \otimes y'$ it means that the rectangle with horizontal side x and vertical side y may be converted to the rectangle with horizontal side x' and vertical side y' in this way.

5.6.1. If x, y, x', y' are rational, and $xy = x'y'$, show that $x \otimes y = x' \otimes y'$.

This prompts the question: are rectangles of equal area always equivalent under vertical and horizontal cut and paste? In particular, can the rectangle $\sqrt{2} \otimes 1/\sqrt{2}$ be converted to the unit square $1 \otimes 1$ in this way? We shall answer this question in the next section, where it is revealed that tensors capture not only equivalence by cutting and pasting, but also the relations between rational and irrational numbers. The following exercise gives another clue to the role of rational numbers.

5.6.2. Show that $rx \otimes y = x \otimes ry$ for any rational r.

5.7* Additive Functions

The nearest thing to a tensor $l_1 \otimes \alpha_1 + l_2 \otimes \alpha_2 + \cdots + l_k \otimes \alpha_k$ we can build in ordinary algebra is a function $l_1 f(\alpha_1) + l_2 f(\alpha_2) + \cdots + l_k f(\alpha_k)$, where f is a function with the properties

$$lf(\alpha + \beta) = lf(\alpha) + lf(\beta)$$
$$(l + m)f(\alpha) = lf(\alpha) + mf(\alpha)$$
$$f(\pi) = 0$$

analogous to Rules 1, 2, and 3 for tensors. The second of these properties is true of all real functions (by the distributive law), so the relevant functions f are actually those with $f(\pi) = 0$ and the property $f(\alpha + \beta) = f(\alpha) + f(\beta)$, called *additivity*.

Admittedly, the only additive functions close at hand are the functions $f(x) = kx$, and the only one of these with $f(\pi) = 0$ is the constant function 0. This does not look promising, but luckily we do not need additive functions defined on all of $\mathbb{R}$. We only need additive functions defined on finite sets of reals, and enough of these can be obtained with the help of the following concept.

Definition A *basis over* $\mathbb{Q}$ for a finite set S of reals is a set $\{\beta_1, \beta_2, \ldots, \beta_n\}$ such that

1. Each x in S is expressible as $x = \beta_1 r_1 + \beta_2 r_2 + \cdots + \beta_n r_n$ for some rationals $r_1, r_2, \ldots, r_n$. (We call x a *rational combination* of $\beta_1, \beta_2, \ldots, \beta_n$).

2. The β_j are *rationally independent*, that is, $\beta_1 r_1 + \beta_2 r_2 + \cdots + \beta_n r_n = 0$ for rationals $r_1, r_2, \ldots, r_n$ only if all $r_j = 0$.

It follows that if $\{\beta_1, \beta_2, \ldots, \beta_n\}$ is a basis over $\mathbb{Q}$ for S then

- Each x in S is *uniquely* expressible in the form $x = \beta_1 r_1 + \beta_2 r_2 + \cdots + \beta_n r_n$ with rationals $r_1, r_2, \ldots, r_n$. Because if

$$x = \beta_1 r_1 + \beta_2 r_2 + \cdots + \beta_n r_n = \beta_1 s_1 + \beta_2 s_2 + \cdots + \beta_n s_n$$

are two different expressions for x with rational coefficients, we have

$$\beta_1 (r_1 - s_1) + \beta_2 (r_2 - s_2) + \cdots + \beta_n (r_n - s_n) = 0$$

with not all the rational coefficients $(r_j - s_j)$ zero, contrary to the rational independence of $\beta_1, \beta_2, \ldots, \beta_n$.

- The function

$$f_i(x) = r_i, \quad \text{where} \quad x = \beta_1 r_1 + \beta_2 r_2 + \cdots + \beta_n r_n$$

is well defined for all x in S and is additive because

$$x' = \beta_1 r'_1 + \beta_2 r'_2 + \cdots + \beta_n r'_n$$
$$\Rightarrow x + x' = \beta_1 (r_1 + r'_1) + \beta_2 (r_2 + r'_2) + \cdots + \beta_n (r_n + r'_n)$$
$$\Rightarrow f_i(x + x') = r_i + r'_i = f_i(x) + f_i(x').$$

Thus a basis $\{\beta_1, \beta_2, \ldots, \beta_n\}$ over $\mathbb{Q}$ gives us functions f_i that are not only additive but are equal to 1 on β_i and 0 on other basis members. Such functions are just what we will need, so we would like a basis over $\mathbb{Q}$ for each finite set.

Construction of bases over $\mathbb{Q}$.   *A finite set of reals has a basis over $\mathbb{Q}$.*

Proof   Suppose $S = \{x_1, x_2, \ldots, x_m\}$ is a finite set of reals. Choose x_1 as the first basis element β_1. Then look at $x_2, x_3, x_4 \ldots$ in turn and let β_2 be the first x_j that is not a rational multiple of β_1, let β_3 be the next x_j that is not a rational combination of β_1 and β_2, and so on.

I claim that the set $\{\beta_1, \beta_2, \ldots, \beta_n\}$ obtained in this way is rationally independent. If not, there are rationals $r_1, r_2, \ldots, r_n$, not all zero, with $\beta_1 r_1 + \beta_2 r_2 + \cdots + \beta_n r_n = 0$. But if r_i is the last of them $\neq 0$ we have $\beta_i = -\beta_1 r_1/r_i - \beta_2 r_2/r_i - \cdots - \beta_{i-1} r_{i-1}/r_i$, contrary to the choice of β_i as some x_j that is not a rational combination of the previously chosen $\beta_1, \beta_2, \ldots$.

Also, each x_j in S is a rational combination of $\beta_1, \beta_2, \dots, \beta_n$, either as a chosen basis element or as a rational combination of previously chosen basis elements. Hence $\{\beta_1, \beta_2, \dots, \beta_n\}$ is a basis for S over $\mathbb{Q}$. □

This gives us enough additive functions. Now we use them to link rational independence with equidecomposability, in the following crucial theorem.

Rational independence theorem. *If $\alpha_1, \alpha_2, \dots, \alpha_k$ and π are rationally independent, $l_1 \otimes \alpha_1 + l_2 \otimes \alpha_2 + \cdots + l_k \otimes \alpha_k = 0$ only if $l_1 = l_2 = \cdots = l_k = 0$.*

Proof By definition of tensors, $l_1 \otimes \alpha_1 + l_2 \otimes \alpha_2 + \cdots + l_k \otimes \alpha_k = 0$ means $l_1 \otimes \alpha_1 + l_2 \otimes \alpha_2 + \cdots + l_k \otimes \alpha_k$ can be converted to 0 by applying the rules

$$l \otimes (\alpha + \beta) = l \otimes \alpha + l \otimes \beta,$$
$$(l + m) \otimes \alpha = l \otimes \alpha + m \otimes \alpha,$$
$$l \otimes \pi = 0.$$

We can similarly convert $l_1 f(\alpha_1) + l_2 f(\alpha_2) + \cdots + l_k f(\alpha_k)$ to 0 for any additive function f with $f(\pi) = 0$, by applying the rules

$$lf(\alpha + \beta) = lf(\alpha) + lf(\beta),$$
$$(l + m)f(\alpha) = lf(\alpha) + mf(\alpha),$$
$$f(\pi) = 0,$$

provided f is defined on the finite set S of angles occurring in the proof that $l_1 \otimes \alpha_1 + l_2 \otimes \alpha_2 + \cdots + l_k \otimes \alpha_k = 0$.

Now if $\alpha_1, \alpha_2, \dots, \alpha_k$ and π are rationally independent, they can be made members of a basis over $\mathbb{Q}$ of any finite set containing them, such as S. For example, put $\alpha_1, \alpha_2, \dots, \alpha_k$ and π first on the list of members of S, and use the preceding construction. We therefore have an additive function f_i on S that is 1 on α_i and 0 on the other members of the basis.

Because f_i is defined on S, that is, on all angles occurring in the proof that $l_1 \otimes \alpha_1 + \cdots + l_k \otimes \alpha_k = 0$, we can similarly prove that $l_1 f_i(\alpha_1) + \cdots + l_k f_i(\alpha_k) = 0$. But, by definition of f_i, the latter equation is simply $l_i = 0$. And because i is arbitrary, this means that $l_1 = l_2 = \cdots = l_k = 0$. □

Exercises

5.7.1. There is a similar but simpler rational independence theorem for the tensor product $\mathbb{R} \otimes \mathbb{R}$ discussed in the previous set of exercises.

5.7.2. Show that if $y_1, y_2, \dots, y_k$ are rationally independent, then $x_1 \otimes y_1 + x_2 \otimes y_2 + \cdots + x_k \otimes y_k = 0$ in $\mathbb{R} \otimes \mathbb{R}$ only if $x_1 = x_2 = \cdots = x_k = 0$.

5.7.3. Deduce from Exercise 5.7.1 that the rectangle with vertical side $\sqrt{2}$ and horizontal side $1/\sqrt{2}$ cannot be converted to the unit square by vertical and horizontal cut and paste.

The same idea may be used to show that the $\sqrt{2} \otimes \sqrt{3}$ rectangle cannot be converted to the $\sqrt{6} \otimes 1$ rectangle except by oblique cutting and pasting.

5.7.3. Show the rational independence of

- $\sqrt{2}$ and $\sqrt{3}$,
- $\sqrt{2}$, $\sqrt{3}$, and $\sqrt{6}$,

and hence conclude that the $\sqrt{2} \otimes \sqrt{3}$ rectangle cannot be converted to the $\sqrt{6} \otimes 1$ rectangle by vertical and horizontal cut and paste.

5.8* The Tetrahedron and the Cube

The rational independence theorem tells us that $l \otimes \alpha \neq 0$ if $l \neq 0$ and α is rationally independent of π, that is, if α is not a rational multiple of π. Now the Dehn invariant of the regular tetrahedron with unit edges is $6 \otimes \alpha$, where $\cos \alpha = 1/3$ by Section 5.5*. We also know that the regular tetrahedron is equidecomposable with a cube only if its Dehn invariant equals the Dehn invariant of the cube, which is 0 by Section 5.6*.

Putting all this together, it remains to prove that the dihedral angle α of the regular tetrahedron is not a rational multiple of π. This is a pleasant exercise using the addition formula for cosine and some elementary number theory.

Dehn's theorem   *The regular tetrahedron is not equidecomposable with the cube.*

Proof If α is a rational multiple of π then $n\alpha = m\pi$ for some integers m and n, in which case $\cos n\alpha = \pm 1$. We shall show that this is impossible for any natural number n (which is sufficient, because if $n\alpha = m\pi$ we can take n to be positive by changing the sign of m if necessary). In fact, we shall use induction on n to prove the stronger statement S_n:

$$\cos n\alpha = \frac{q_n}{3^n} \quad \text{for some integer } q_n \text{ not divisible by 3.}$$

S_1 is true because $\cos \alpha = 1/3$. Now suppose $S_1, S_2, \ldots, S_k$ are all true. We prove S_{k+1} by means of the identity

$$\cos(k+1)\alpha + \cos(k-1)\alpha = 2\cos k\alpha \cos \alpha,$$

which comes from adding the two addition formulas

$$\cos(k+1)\alpha = \cos k\alpha \cos \alpha - \sin k\alpha \sin \alpha,$$
$$\cos(k-1)\alpha = \cos k\alpha \cos \alpha + \sin k\alpha \sin \alpha.$$

The identity says that

$$\cos(k+1)\alpha = 2\cos k\alpha \cos \alpha - \cos(k-1)\alpha,$$

and by our induction hypothesis we have integers q_k and q_{k-1}, not divisible by 3, such that

$$\cos k\alpha = \frac{q_k}{3^k} \quad \text{and} \quad \cos(k-1)\alpha = \frac{q_{k-1}}{3^{k-1}}.$$

Because $\cos \alpha = 1/3$, it follows that

$$\cos(k+1)\alpha = \frac{(2/3)q_k}{3^k} - \frac{q_{k-1}}{3^{k-1}} = \frac{2q_k - 9q_{k-1}}{3^{k+1}} = \frac{q_{k+1}}{3^{k+1}},$$

where $q_{k+1} = 2q_k - 9q_{k-1}$ is also not divisible by 3 because $9q_{k-1}$ is and $2q_k$ is not.

This completes the induction step, and hence $\cos n\alpha \neq \pm 1$, for all natural numbers n, as required. $\square$

Exercises

The proof of Dehn's theorem can be generalized to show that a rational multiple of π has irrational cosine, except when the angle is one of $\pi/3$,

$\pi/2$, π, or their integer multiples. The proof of this can be broken into a few easy stages. The first is to check what happens with $\pi/3$, $\pi/2$, and π.

5.8.1. Show that $\cos n\alpha$ is rational when $\alpha = \pi/3$, $\pi/2$, or π and n is an integer. Also show that the values of $\cos n\alpha$ in these cases are 0, ± 1, and $\pm\frac{1}{2}$.

Perhaps the fraction $\frac{1}{2} = \cos \pi/3$ is an exceptional value of $\cos \alpha$. What about fractions of the form $s/2^t$?

5.8.2. Suppose $\cos \alpha = u/2^v$, where u is an odd integer and v is an integer ≥ 2. Show by induction on n that

$$\cos n\alpha = \frac{u_n}{2^{nv-n+1}}, \quad \text{where } u_n \text{ is an odd integer.}$$

Hence deduce that α is not a rational multiple of π.

This disposes of the rational values of $\cos \alpha$ whose denominator is a power of 2. From now on, we can assume that $\cos \alpha$ has a denominator divisible by an odd prime p, so $\cos \alpha = r/p^v$, where $r = s/t$, s and t are integers not divisible by p, and v is an integer ≥ 1.

5.8.3. If $\cos \alpha = r/p^v$, with r and v as just described, show by induction on n that

$$\cos n\alpha = \frac{r_n}{p^{nv}},$$

where $r_n = s_n/t_n$, and s_n and t_n are integers not divisible by p. Hence deduce that α is not a rational multiple of π.

These results have interesting implications for Pythagorean triples and rational points on the unit circle. The Pythagorean triple (a, b, c) represents a triangle whose angle α (between the sides a and c) has rational cosine a/c. It follows from Exercises 5.8.1, 5.8.2, and 5.8.3 that α is not a rational multiple of π unless $\cos \alpha = \pm 1/2$, and in fact this is also impossible.

5.8.4. We cannot have $a/c = 1/2$ in a Pythagorean triple. Why?

5.8.5. Deduce that division of the unit circle into equal parts by rational points is possible only when the number of parts is 2 or 4.

5.9   Discussion

Formulas for π

Finding the value of π, the circumference of the circle of diameter 1, is one of the oldest and most fundamental problems in mathematics. Because the circle is the simplest curve, apart from the straight line, finding the length of the circle is surely the most obvious question in geometry once the basic questions about lines have been answered, as they are by the Pythagorean theorem. How surprising, then, that finding the length of the circle is *nothing* like finding the length of a line!

The Greeks were baffled by the problem and could only find approximations such as the one found by Archimedes,

$$3\frac{10}{71} < \pi < 3\frac{1}{7}.$$

The better approximation 355/113 (accurate to six decimal places) was found by the Chinese mathematician Zǔ Chōngzhi (429–500 A.D.) and was later rediscovered in Europe, along with approximations to more and more decimal places. However, finite rational approximations give little insight into the nature of π, because π is irrational. One would prefer an exact infinite description, provided it yields approximations in a uniform and comprehensible way.

Such a description, infinite yet miraculously simple, was first found in India around 1500 A.D. It expresses $\pi/4$ as an infinite sum of rational numbers:

$$\frac{\pi}{4} = 1 - \frac{1}{3} + \frac{1}{5} - \frac{1}{7} + \cdots.$$

Like the Pythagorean theorem, this formula is one of the universal treasures of mathematics, which one might expect to be discovered by any advanced civilization. It was rediscovered in Europe around 1670, with a similar proof. One of its discoverers, Gottfried Wilhelm Leibniz, was so enchanted by the simplicity of the formula that he declared: "God loves odd numbers."

The original discoverer of the formula is not known for certain: the earliest surviving proof (around 1530) credits Nīlakaṇṭha, who flourished around 1500, but a slightly later manuscript by

Jyesthadeva credits Madhava (1340–1425). The proof given by Jyesthadeva is based on a geometric lemma and a limit calculation involving sums of powers of integers. Actually, a more general result is proved, which we would call the infinite series for the inverse tan function:

$$\tan^{-1} x = x - \frac{x^3}{3} + \frac{x^5}{5} - \frac{x^7}{7} + \cdots .$$

The formula for π results by substituting $x = 1$, because $\tan^{-1} 1 = \pi/4$. A reconstruction of the proof may be found in Katz (1993), pp. 452–453.

The formula was rediscovered in Europe by James Gregory and Leibniz around 1670, using calculus. For readers familiar with calculus, it should be mentioned that the fundamental theorem of calculus makes a dramatic simplification in the Indian proof, replacing the awkward problem of evaluating

$$\lim_{n \to \infty} \frac{1^{2i} + 2^{2i} + \cdots + (n-1)^{2i}}{n^{2i+1}}$$

by the integration of x^{2i}, which every beginner in calculus can do. Somewhat earlier, before the fundamental theorem of calculus was known, Wallis (1655) used some ingenious guesswork to discover an expression for π as an *infinite product*:

$$\frac{\pi}{4} = \frac{2}{3} \cdot \frac{4}{3} \cdot \frac{4}{5} \cdot \frac{6}{5} \cdot \frac{6}{7} \cdots .$$

This result can be obtained rigorously by the technique of integration by parts and is now a common exercise in calculus textbooks. Wallis' colleague Brouncker managed to transform the infinite product into the infinite *continued fraction*

$$\frac{4}{\pi} = 1 + \cfrac{1^2}{2 + \cfrac{3^2}{2 + \cfrac{5^2}{2 + \cfrac{7^2}{2 + \ddots}}}},$$

and this result was also reported by Wallis (1655). (We'll say more about continued fractions in Chapter 8, because they are also of great interest in the study of square roots.)

Brouncker's continued fraction is not of the standard type, which has all the numerators equal to 1, and in fact the standard continued fraction,

$$\pi = 3 + \cfrac{1}{7 + \cfrac{1}{15 + \cfrac{1}{1 + \cfrac{1}{292 + \ddots}}}},$$

does not have any discernible pattern of denominators. It is nevertheless an interesting curiosity that the truncated fraction,

$$3 + \cfrac{1}{7 + \cfrac{1}{15 + \frac{1}{1}}},$$

is precisely Zǔ Chōngzhī's approximation to π, 355/113. The exceptional accuracy of this approximation is partly due to stopping just before the large denominator 292.

Euler (1748) linked Brouncker's continued fraction, and hence Wallis's infinite product, to the Indian series for $\pi/4$ by transforming the series into the continued fraction. Thus Wallis' product is in some sense a rediscovery of the formula

$$\frac{\pi}{4} = 1 - \frac{1}{3} + \frac{1}{5} - \frac{1}{7} + \cdots,$$

or at least confirmation of its fundamental nature.

Euler (1748) also found a whole family of formulas for even powers of π, starting with

$$\frac{\pi^2}{6} = \frac{1}{1^2} + \frac{1}{2^2} + \frac{1}{3^2} + \frac{1}{4^2} + \cdots$$

$$\frac{\pi^4}{90} = \frac{1}{1^4} + \frac{1}{2^4} + \frac{1}{3^4} + \frac{1}{4^4} + \cdots$$

$$\frac{\pi^6}{945} = \frac{1}{1^6} + \frac{1}{2^6} + \frac{1}{3^6} + \frac{1}{4^6} + \cdots.$$

He transformed these into infinite products involving the prime numbers, for example,

$$\frac{\pi^2}{6} = \frac{1}{1 - 2^{-2}} \cdot \frac{1}{1 - 3^{-2}} \cdot \frac{1}{1 - 7^{-2}} \cdot \frac{1}{1 - 11^{-2}} \cdots,$$

using the wonderful *Euler product formula*

$$\frac{1}{1^s} + \frac{1}{2^s} + \frac{1}{3^s} + \frac{1}{4^s} + \cdots = \frac{1}{1 - 2^{-s}} \cdot \frac{1}{1 - 3^{-s}} \cdot \frac{1}{1 - 7^{-s}} \cdot \frac{1}{1 - 11^{-s}} \cdots,$$

which is valid for any $s > 1$, and is essentially equivalent to unique prime factorization.

(This last remark is supposed to tempt you to look for an explanation of the Euler product formula. You can find it by expanding $\frac{1}{1-p^{-1}}$ as a geometric series

$$\frac{1}{1 - p^{-s}} = 1 + p^{-s} + p^{-2s} + p^{-3s} + \cdots.$$

Then observe that the product of these series, for all primes p, includes each term n^{-s} exactly once, because each $n > 1$ equals exactly one product of primes. This formula shows, incidentally, that mathematicians were aware of unique prime factorization well before it was explicitly stated by Gauss in 1801.)

Additive Functions and the Axiom of Choice

The construction of additive functions in Section 5.7* is tailored to solve Hilbert's third problem and no more; it sidesteps the awkward question: *is there a nonconstant additive function f, defined on all of $\mathbb{R}$, such that $f(\pi) = 0$?* However, this is a very interesting question. We shall therefore explore it a little further, if only because a *yes* answer could simplify the solution of Hilbert's third problem.

Suppose that $f(\pi) = 0$ and f is additive, that is, $f(\alpha+\beta) = f(\alpha)+f(\beta)$ for all real numbers α and β. It follows that $f(\pi/2) = 0$ also, because

$$0 = f(\pi) = f(\pi/2) + f(\pi/2) = 2f(\pi/2).$$

It similarly follows that $f(\pi/n) = 0$ for any positive integer n, and hence

$$f(m\pi/n) = mf(\pi/n) = 0$$

for any integer m. Thus in fact $f(r\pi) = 0$ for any rational number r. But the number $r\pi$ can be made arbitrarily close to any real number we please, by suitable choice of the rational number r. Hence, if f is a continuous function, it can only be the constant function 0.

Thus we must certainly give up the idea of looking for continuous additive functions, and we cannot avoid f being zero at all rational multiples $r\pi$ of π. Fortunately, there are many numbers outside this set, for example, the nonzero rationals. We can choose $f(a_1)$ to be any value we like, say, 1, when a_1 is not of the form $r\pi$ for a rational r. The value of $f(a_1)$ determines the value $f(r'a_1) = r'f(a_1)$ for any rational r', and hence (by additivity again) the value of $f(r\pi + r'a_1)$ for any pair of rationals r, r'.

Now there are countably many pairs of rationals (by an argument like that used in Section 3.10 to prove there are countably many fractions), and hence countably many numbers $r\pi + r'a_1$. Numbers exist outside this set, by the uncountability of $\mathbb{R}$ proved in Section 3.10, so we are still free to choose the value of f on any one of them, say, a_2. This determines the value of f on the countably many numbers $r\pi + r'a_1 + r''a_2$ where r, r', and r'' are rational, and so on. As long as the values on which f is defined do not exhaust $\mathbb{R}$, we can choose a number a on which f is undefined, and give $f(a)$ any value we like.

The trouble is, doing this for infinitely many values $a_1, a_2, a_3, \ldots$ will *not* exhaust $\mathbb{R}$, precisely because $\mathbb{R}$ is uncountable. Our intuition balks at continuing the sequence of choices to uncountable length, and the best we can do is *assume* that it is possible. This assumption (in a more precise form, of course) is called the *axiom of choice*.

No matter how we approach the problem of additive functions on $\mathbb{R}$, the axiom of choice is needed. For example, we could approach the problem as we did for finite additive functions in Section 5.7*, by constructing a *basis over* $\mathbb{Q}$ for $\mathbb{R}$. Choose any real β_1 to be the first member of the basis, and more generally let β_{n+1} be any real that is not one of the countably many rational combinations of $\beta_1, \beta_2, \ldots, \beta_n$. The axiom of choice allows us to assume that $\mathbb{R}$ can be exhausted by extending the sequence of choices to uncountable length. If so, we then have a basis over $\mathbb{Q}$ for $\mathbb{R}$, and each real number x is a unique rational combination of basis members.

If β_i is any basis member, it follows as in Section 5.7* that the following function f_i is additive.

$$f_i(x) = \text{coefficient of } \beta_i \text{ in the expression for } x$$
$$\text{as a rational combination of basis elements.}$$

We can therefore use the functions f_i as before to solve Hilbert's third problem. This argument is more in the mathematical mainstream, because it is quite usual to assume the existence of bases.

On the other hand, one does not want to assume the axiom of choice unnecessarily, because it is a dubious axiom.

In fact, the situation of the axiom of choice is not unlike the situation of Euclid's parallel axiom before the discovery of non-Euclidean geometry. The axiom of choice is not as natural as the other axioms of set theory, and we know that it can be neither proved nor disproved from them. But the axioms of set theory are the most powerful axioms we know to settle mathematical questions; anything outside them, such as the axiom of choice, is currently a matter of blind faith. From this point of view, it is comforting that Hilbert's third problem can be solved without it.

6 Finite Arithmetic

6.1 Three Examples

This book began by stressing the role of infinity in mathematics, its presence in the concept of number, and the importance of learning to live with it. Since then, infinity has appeared in many situations, and we have seen many ways to approach and tame it. Still, it is remarkable how often we succeed. Even if the world of ideas is infinite, as Dedekind believed, there is no doubt that *proofs* are finite, so success with infinity depends on capturing its properties by finite methods. Induction is one such method, but there are others.

Several times we have proved results about all numbers by considering only a finite set, such as the set of remainders that can occur when an integer is divided by 2 or 4. Apparently, the infinitude of the set of integers is irrelevant in some problems, and a way to see the relevant part is to focus on remainders. If so, the arithmetic of remainders deserves further clarification and development.

In daily life, we know it can be meaningful and useful to do arithmetic with remainders. For example, we add 3 hours to 11 o'clock and get 2 o'clock, the remainder when $11 + 3$ is divided by 12. Addition *mod 12*, as this is called, is the ideal arithmetic for keeping the time of day and is no great mathematical challenge. The plot

177

thickens when we combine addition and multiplication on finite sets. The idea not only seems to work, it actually seems capable of producing serious results, which are hard to notice or understand in the infinite set of natural numbers.

Example 1.   Even and odd.

In proving that $2n^2 \neq m^2$ for all integers m and n (Section 1.1) we used the facts that even × even = even and odd × odd = odd. These facts follow from facts about 0 and 1: that if two integers leave remainder 0 on division by 2 then so does their product, and if two integers leave remainder 1 on division by 2 then so does their product. Such facts, and others such as even + even = even and even + odd = odd, hint at an "arithmetic of even and odd" that reflects the behavior of 0 and 1 as remainders on division by 2.

Example 2.   Sums of two squares.

In proving that a primitive Pythagorean triple (a, b, c) cannot have a and b both odd (Section 4.2, Exercises), we used the fact that a square leaves remainder 0 or 1 on division by 4. Because this implies that the sum of odd squares leaves remainder 2, the sum of odd squares cannot be a square. Again it looks like there is a finite arithmetic in the background here, this time an arithmetic of the remainders 0, 1, 2, 3 on division by 4.

Example 3.   Rational points on $x^2 + y^2 = 3$.

In Section 4.4 we observed that these exist only if there are relatively prime integers a, b, and c such that $a^2 + b^2 = 3c^2$. We found that such integers do not exist by a similar appeal to remainders on division by 4: a and b are not both even, so at least one of them leaves remainder 1 on division by 4, and so does its square. Hence $a^2 + b^2$ leaves remainder 1 or 2, whereas $3c^2$ leaves remainder 0 or 3.

These examples seem to be telling us that it is useful to divide the integers into finitely many "classes" according to their remainders on division by n. Certain things are impossible in the integers merely because they are impossible in a suitably chosen set of remainders, so a fruitless infinite search through the integers may be avoided by looking instead through a finite set.

Exercises

Our ordinary base 10 system of numerals is quite convenient for finding remainders on division by 2, 4, 8, ... To find the remainder of any number on division by 2, take the remainder of its last digit; to find the remainder on division by 4, take the remainder of its last two digits, treating them as a two-digit number, and so on.

6.1.1. Show that the remainder on division by 8 can be found as the remainder of the last three digits, regarded as a three-digit number.

6.1.2. Explain why the last n digits suffice to find the remainder of any number on division by 2^n.

Another problem that can be settled by looking at remainders is Exercise 4.6.5*. A related problem is the following property of sums of three squares.

6.1.3. If x, y, and z are integers, show that $x^2 + y^2 + z^2$ leaves remainder 0, 1, 2, 3, 4, 5, or 6 on division by 8.

6.1.4. Deduce from Exercise 6.1.3 that a number of the form $8n + 7$, for any integer n, is not a sum of three squares.

Thus there are infinitely many natural numbers that are not sums of three squares. However, every natural number is a sum of four squares, by a famous theorem of Lagrange (1770). A few years later, Legendre found that the natural numbers that are not sums of three squares are those of the form $4^m(8n+7)$. Sums of two squares are also interesting, and we shall say more about them in this chapter and the next. The first thing to know about them is the following, which can be proved by considering remainders on division by 4.

6.1.5. Show that an integer of the form $4n+3$ is not a sum of two squares.

6.2 Arithmetic mod n

The arithmetic of remainders on division by n was first made precise by Gauss in his famous book the *Disquisitiones Arithmeticae* of 1801. Gauss based this arithmetic on the idea of *congruence mod n*, for which he introduced the notation

$$a \equiv b \pmod{n}.$$

This expression is read "a is congruent to b modulo (or simply mod) n" and it means that a and b leave the same remainder on division by n. Putting it more concisely, $a \equiv b \pmod{n}$ means that n divides $a - b$. The natural number n is called the *modulus*.

It is sometimes convenient to use the notation $a \bmod n$ for the remainder when a is divided by n. Then the congruence $a \equiv b \pmod{n}$ can be written as the ordinary equation $a \bmod n = b \bmod n$.

We are already familiar with the concept of congruence when the modulus $n = 2$. It is just another way to describe the even and odd numbers. The even numbers are those congruent to 0 (mod 2), and the odd numbers are those congruent to 1 (mod 2).

Numbers that are congruent mod n are interchangeable in some remainder calculations. For example, it is valid to say things like "odd + even = odd," "odd − even = odd," and "odd × even = even" because adding, subtracting, or multiplying *any* odd number and *any* even number gives a result with the same remainder on division by 2. The situation is similar with any modulus n in place of 2: numbers that are congruent mod n are *arithmetically equivalent* in the sense that they produce the same results in sums, differences, and products.

Arithmetic equivalence mod n.   *If $a_1 \equiv a_2 \pmod{n}$ and $b_1 \equiv b_2 \pmod{n}$, then*

$$a_1 + b_1 \equiv a_2 + b_2 \pmod{n}$$
$$a_1 - b_1 \equiv a_2 - b_2 \pmod{n}$$
$$a_1 b_1 \equiv a_2 b_2 \pmod{n}.$$

Proof   Because $c \equiv d \pmod{n}$ means n divides $c - d$, the two given congruences can be translated into statements about divisibility. Manipulating them slightly and translating back gives the three required congruences quite easily, especially the first two. The first goes like this:

$$a_1 \equiv a_2 \pmod{n} \text{ and } b_1 \equiv b_2 \pmod{n}$$
$$\Rightarrow \quad n \text{ divides } a_1 - a_2 \text{ and } n \text{ divides } b_1 - b_2$$

$$\Rightarrow \quad n \text{ divides } (a_1 - a_2) + (b_1 - b_2)$$
$$\Rightarrow \quad n \text{ divides } (a_1 + b_1) - (a_2 + b_2)$$
$$\Rightarrow \quad a_1 + b_1 \equiv a_2 + b_2 \pmod{n}.$$

The second is the same, except for suitable replacement of $+$ signs by $-$ signs.

For the third, the expression we want n to divide, namely, $a_1b_1 - a_2b_2$, must be written in terms of the expressions we know are divisible by n, namely $a_1 - a_2$ and $b_1 - b_2$. Some experimentation gives

$$a_1b_1 - a_2b_2 = a_1(b_1 - b_2) + b_2(a_1 - a_2),$$

so a proof of the third congruence is:

$$a_1 \equiv b_1 \pmod{n} \text{ and } a_2 \equiv b_2 \pmod{n}$$
$$\Rightarrow \quad n \text{ divides } a_1 - a_2 \text{ and } n \text{ divides } b_1 - b_2$$
$$\Rightarrow \quad n \text{ divides } a_1(b_1 - b_2) + b_2(a_1 - a_2)$$
$$\Rightarrow \quad n \text{ divides } a_1b_1 - a_2b_2$$
$$\Rightarrow \quad a_1b_1 \equiv a_2b_2 \pmod{n}. \qquad \qquad \square$$

The arithmetic equivalence of congruent numbers means that some common manipulations with equations are also valid for congruences.

1. We can add them: if $a_1 \equiv a_2 \pmod{n}$ and $b_1 \equiv b_2 \pmod{n}$ then $a_1 + b_1 \equiv a_2 + b_2 \pmod{n}$.

2. We can subtract them: if $a_1 \equiv a_2 \pmod{n}$ and $b_1 \equiv b_2 \pmod{n}$ then $a_1 - b_1 \equiv a_2 - b_2 \pmod{n}$.

3. We can multiply them: if $a_1 \equiv a_2 \pmod{n}$ and $b_1 \equiv b_2 \pmod{n}$ then $a_1b_1 \equiv a_2b_2 \pmod{n}$.

Exercises

An old rule of arithmetic, called *casting out nines*, has a nice explanation in terms of arithmetic mod 9. The rule says that a number is divisible by 9 if the sum of its (base 10) digits is divisible by 9. For example, 774 is divisible by 9 because $7 + 7 + 4 = 18$ is divisible by 9.

6.2.1. AM radio frequencies in Melbourne are 621, 693, 774, 855, 927, 1116, 1224, 1278, 1377, 1422, 1503, 1593 (kHz). What do you notice about these numbers?

Casting out nines is easily understood when one recalls how base 10 numerals are built from their digits and powers of 10.

6.2.2. A base 10 numeral $a_k a_{k-1} \ldots a_1 a_0$ stands for $a_k 10^k + a_{k-1} 10^{k-1} + \cdots + a_1 10 + a_0$. Explain.

6.2.3. Notice that $10 \equiv 1 \pmod 9$, and hence

$$10 \times 10 \equiv 1 \times 1 \pmod 9, \quad \ldots, \quad 10^k \equiv 1^k \pmod 9.$$

Deduce from Exercise 6.2.2 that

$$a_k a_{k-1} \ldots a_1 a_0 \equiv a_k + a_{k-1} + \cdots + a_1 + a_0 \pmod 9.$$

Thus $a_k a_{k-1} \ldots a_1 a_0$ is divisible by 9 if and only if $a_k + a_{k-1} + \cdots + a_1 + a_0$ is; in fact they both have the same remainder on division by 9.

6.2.4. Notice also that $10 \equiv 1 \pmod 3$, and hence show similarly that

$$a_k a_{k-1} \ldots a_1 a_0 \equiv a_k + a_{k-1} + \cdots + a_1 + a_0 \pmod 3.$$

This gives a test for divisibility by 3 by "casting out threes." The next simplest is a test for divisibility by 11. Here again one needs to sum the digits, but now *taken with alternate + and − signs*. For example, 11 divides 16577, because 11 divides $1 - 6 + 5 - 7 + 7 = 0$.

6.2.5. Use the fact that $10 \equiv -1 \pmod{11}$ to show that

$$a_k a_{k-1} \ldots a_1 a_0 \equiv (-1)^k a_k + (-1)^{k-1} a_{k-1} + \cdots - a_1 + a_0 \pmod{11}.$$

Of course it is easy to find powers of 10 when 10 is congruent to 1 or −1. But in fact powers of any number are quite easy to evaluate, modulo any n. We need work only with remainders, and therefore we only need to multiply numbers $< n$. Also, large powers can be reached quickly by squaring wherever possible.

For example, to find large powers of 3, mod 19, we evaluate $3^2, 3^4, 3^8, 3^{16}, \ldots$ in succession by repeatedly finding the remainder on division by 19 and squaring it. The first step that involves a genuine remainder is where $3^4 = 81 \equiv 5 \pmod{19}$, and therefore $3^8 \equiv 5^2 \equiv 25 \equiv 6 \pmod{19}$. When enough powers $3^2, 3^4, 3^8, 3^{16}, \ldots$ have been found, other powers can be found as products of them. For example, $3^{100} = 3^{64} 3^{32} 3^4$ because $100 = 64 + 32 + 4$.

6.2.6. Find 3^{16}, 3^{32}, and 3^{64} mod 19, and show $3^{100} \equiv 16 \pmod{19}$.

6.2.7.* The method of repeated squaring depends on the fact that every natural number is a sum of powers of 2. Explain the dependence, and explain why the fact is true.

(Hint: Subtract the largest power of 2 and use descent.)

6.3 The Ring $\mathbb{Z}/n\mathbb{Z}$

The numbers congruent to a given integer a modulo n form the set

$$\{a + nk : k \in \mathbb{Z}\} = \{\ldots, a - 2n, a - n, a, a + n, a + 2n, \ldots\}.$$

We call this set the *congruence class* of a, mod n, and denote it by $a + n\mathbb{Z}$ for short. In particular, the set of all multiples of n,

$$n\mathbb{Z} = \{nk : k \in \mathbb{Z}\} = \{\ldots, -2n, -n, 0, n, 2n, \ldots\},$$

is the congruence class of 0. There are n different congruence classes, one for each remainder on division by n.

For example, there are two congruence classes mod 2:

$$2\mathbb{Z} = \{2k : k \in \mathbb{Z}\} = \{\text{even integers}\}$$

and

$$1 + 2\mathbb{Z} = \{1 + 2k : k \in \mathbb{Z}\} = \{\text{odd integers}\}.$$

We can now give a precise meaning to equations such as "odd + even = odd," "odd − even = odd" and "odd × even = even" by defining the sum, difference, and product of congruence classes.

For any modulus, the definitions say that

$$(\text{class of } a) + (\text{class of } b) = \text{class of } (a + b)$$
$$(\text{class of } a) - (\text{class of } b) = \text{class of } (a - b)$$
$$(\text{class of } a) \times (\text{class of } b) = \text{class of } ab$$

or, in the notation we have just introduced,

$$(a + n\mathbb{Z}) + (b + n\mathbb{Z}) = (a + b) + n\mathbb{Z}$$
$$(a + n\mathbb{Z}) - (b + n\mathbb{Z}) = (a - b) + n\mathbb{Z}$$
$$(a + n\mathbb{Z}) \times (b + n\mathbb{Z}) = ab + n\mathbb{Z}.$$

The first of these should be compared with the very similar definition of the sum of angles in Section 5.2. There, and here, the class of a plus the class of b is the class of $a + b$.

To be careful, we should check that the class of $a + b$ is *well defined*, that it does not depend on the numbers we choose to represent the class of a and the class of b. Suppose on one occasion we take a_1 from the first class and b_1 from the second and we form the class of $a_1 + b_1$. If, on another occasion we take a_2 from the first class and b_2 from the second and we form the class of $a_2 + b_2$, is this the same as the class of $a_1 + b_1$? Yes! In fact, this is precisely the first property of arithmetic equivalence, proved in the previous section: if $a_1 \equiv a_2$ (mod n) and $b_1 \equiv b_2$ (mod n), then $a_1 + b_1 \equiv a_2 + b_2$ (mod n).

Similarly, the difference and product of congruence classes are well defined by the second and third properties of arithmetic equivalence.

We use the symbols $+$, $-$, and $\times$ (or juxtaposition) for sum, difference, and product of congruence classes because they have the same properties as ordinary $+$, $-$, and $\times$. In fact, all the ring properties of $+$, $-$, and $\times$ on $\mathbb{Z}$ (Section 1.4) are "inherited" by the operations on congruence classes.

Here is how the $+$ on congruence classes inherits commutativity from ordinary $+$ on $\mathbb{Z}$:

$$(a + n\mathbb{Z}) + (b + n\mathbb{Z}) = (a + b) + n\mathbb{Z}$$

by definition of $+$ for congruence classes

$$= (b + a) + n\mathbb{Z}$$

by commutativity of $+$ for $\mathbb{Z}$

$$= (b + n\mathbb{Z}) + (a + n\mathbb{Z})$$

by definition of $+$ for congruence classes.

It is equally easy to check that all the ring properties of $\mathbb{Z}$ are inherited by congruence classes; hence *the set of congruence classes* mod n *is a ring.* We denote this ring by $\mathbb{Z}/n\mathbb{Z}$. Informally speaking, $\mathbb{Z}/n\mathbb{Z}$ is what $\mathbb{Z}$ becomes when we pretend that $n = 0$.

Putting it a little more formally, $\mathbb{Z}/n\mathbb{Z}$ is what $\mathbb{Z}$ looks like when we focus on remainders mod n. Arithmetic equivalence mod n allows us to ignore all multiples of n and consistently replace each integer by its remainder. As we saw at the beginning of this chapter, this is often the way to avoid being confused by the irrelevant vastness of $\mathbb{Z}$.

Exercises

The question whether an operation is well defined actually arises in around fifth grade, though you were probably not asked to worry about it then. The product of fractions is defined by

$$\frac{a}{b} \times \frac{c}{d} = \frac{ac}{bd},$$

but what we really want is the product of rational numbers, and a rational number is an infinite set of fractions. For example, what we call the "rational number $\frac{1}{2}$" is really the set

$$\left\{ \frac{1}{2}, \frac{-1}{-2}, \frac{2}{4}, \frac{-2}{-4}, \frac{3}{6}, \frac{-3}{-6}, \cdots \right\}$$

of fractions $k/2k$ for all nonzero integers k. To make sure the product of rationals is well defined, we have to check that the fractions ka/kb and lc/ld give the same product rational for any nonzero integers k and l, and of course they do.

6.3.1. Check that the sum of fractions $\frac{a}{b} + \frac{c}{d} = \frac{ad+bc}{bd}$ is well defined for rationals.

6.3.2. Also check that $\frac{a+c}{b+d}$ is *not* a well-defined "sum" of $\frac{a}{b}$ and $\frac{c}{d}$.

Inheritance of ring properties from $\mathbb{Z}$ implies that if an equation involving $+$, $-$, and $\times$ has a solution $x, y, \ldots$ in $\mathbb{Z}$ then it has a solution in any $\mathbb{Z}/n\mathbb{Z}$. In fact, the solution in $\mathbb{Z}/n\mathbb{Z}$ comes from replacing $x, y, \ldots$ by their congruence classes mod n. This sometimes enables us to prove that an equation has no solution in $\mathbb{Z}$ by showing that it has no solution in a suitably chosen $\mathbb{Z}/n\mathbb{Z}$. This is the basis of the idea, described in Section 6.1, that certain things are impossible in the integers merely because they are impossible in a certain set of remainders.

For example, if the equation $x^2 + y^2 = 4n + 3$ has a solution in $\mathbb{Z}$ then it has a solution in $\mathbb{Z}/4\mathbb{Z}$, where it becomes $x^2 + y^2 = 3$, because $4n + 3 \equiv 3 \pmod{4}$. But we can see whether the equation $x^2 + y^2 = 3$ has any solutions in $\mathbb{Z}/4\mathbb{Z}$ simply by trying the four possible values 0, 1, 2, and 3 for x and y.

6.3.3. Try this, and compare what happens with your previous solution (to Exercise 6.1.5) in terms of remainders.

There are some similar theorems about numbers of the form $x^2 + 2y^2$ and $x^2 + 3y^2$, and their possible remainders on division by 8 and 3, respectively.

6.3.4. Show that the equation $x^2 + 2y^2 = 8n + 5$ has no integer solution by considering it in $\mathbb{Z}/8\mathbb{Z}$. Discuss what happens with other numbers in place of 5.

6.3.5. Show that the equation $x^2 + 3y^2 = 3n + 2$ has no integer solution.

6.4   Inverses mod n

We have not said anything about division mod n so far, with good reason: it doesn't always work. In particular, if

$$ab \equiv ac \ (\mathrm{mod}\ n) \quad \text{and} \quad a \not\equiv 0 \ (\mathrm{mod}\ n)$$

it is not necessarily true that $b \equiv c \ (\mathrm{mod}\ n)$. An example is

$$2 \times 1 \equiv 2 \times 3 \equiv 2 \ (\mathrm{mod}\ 4) \quad \text{but} \quad 1 \not\equiv 3 \ (\mathrm{mod}\ 4).$$

The reason that division by 2 does not work, mod 4, is that 2 does not have an *inverse* mod 4. There is no number m such that $2m \equiv 1$ (mod 4), as can be seen by trying $m = 1, 2, 3$. The numbers 1 and 3 do have inverses mod 4. In fact each is its own inverse: $1 \times 1 \equiv 1$ (mod 4) and $3 \times 3 = 9 \equiv 1$ (mod 4). This means that division by 1 and 3 are valid mod 4. For example, from

$$3 \times b \equiv 3 \times c \ (\mathrm{mod}\ 4)$$

we can conclude that

$$3 \times 3 \times b \equiv 3 \times 3 \times c \ (\mathrm{mod}\ 4),$$

multiplying both sides by 3. This is the same as

$$b \equiv c \pmod{4}$$

because $3 \times 3 \equiv 1 \pmod{4}$.

In general, division by a in mod n arithmetic is possible precisely when a has an *inverse* mod n, a number m such that $am \equiv 1 \pmod{n}$. It is therefore a question of knowing which numbers have inverses mod n, and this question has a very neat answer.

Criterion for inverses mod n.   *The number a has an inverse mod n if and only if $\gcd(a, n) = 1$.*

Proof   To show this proof concisely, we use the symbol $\Leftrightarrow$ for "if and only if."

$$a \text{ has an inverse mod } n \Leftrightarrow am \equiv 1 \pmod{n} \text{ for some integer } m$$
$$\Leftrightarrow n \text{ divides } am - 1$$
$$\Leftrightarrow am + nl = 1 \text{ for some integers } l \text{ and } m$$
$$\Leftrightarrow \gcd(a, n) = 1.$$

The last $\Leftrightarrow$ follows from the results about the gcd in Section 1.5.

$$am + nl = 1 \Rightarrow \gcd(a, n) = 1$$

because any divisor of a and n divides the left hand side, and hence divides 1;

$$\gcd(a, n) = 1 \Rightarrow am + nl = 1 \text{ for some integers } l \text{ and } m$$

because $\gcd(a, n) = am + nl$ for integers l and m.   □

This criterion gives a more general explanation why 2 has no inverse in arithmetic mod 4. It cannot have an inverse because $\gcd(2,4) = 2$. We also see that nonzero numbers without inverses will occur for any nonprime modulus n.

But if we have a *prime* modulus p the condition $a \not\equiv 0 \pmod{p}$ means $\gcd(a, p) = 1$, because the only numbers having a larger common divisor with p are the multiples of p, that is, the numbers $\equiv 0 \pmod{p}$. Translating this into the language of congruence classes, as in the previous section, we find: *in the ring $\mathbb{Z}/p\mathbb{Z}$, for prime p, every nonzero element has an inverse*. Now recall from Section 1.4 that a ring in which every nonzero element has an inverse is a field, and we can conclude that *$\mathbb{Z}/p\mathbb{Z}$ is a field.*

This means that some familiar arguments about numbers can also be applied to $\mathbb{Z}/p\mathbb{Z}$. For example, in ordinary algebra we have a theorem that at most n different numbers can satisfy a polynomial equation of degree n (Exercise 5.2.3). Applying the same argument to $\mathbb{Z}/p\mathbb{Z}$ gives the following.

Lagrange's polynomial theorem.   *If $P(x)$ is a polynomial of degree n, then the congruence $P(x) \equiv 0 \pmod p$ has at most n solutions mod p.*

Proof   Suppose $P(x) = a_n x^n + a_{n-1} x^{n-1} + \cdots + a_1 x_1 + a_0$. Because

$$x^k - a^k = (x-a)(x^{k-1} + ax^{k-2} + \cdots + a^{k-2}x + a^{k-1})$$

(as can be checked by multiplying out the right-hand side), it follows that $(x - a)$ is a factor of

$$P(x) - P(a) = a_n(x^n - a^n) + a_{n-1}(x^{n-1} - a^{n-1}) + \cdots + a_1(x - a).$$

Thus

$$P(x) - P(a) = (x-a)Q(x) \quad \text{for some polynomial } Q(x) \text{ of degree } n - 1.$$

Then if $P(a) \equiv 0 \pmod p$ we have

$$P(x) \equiv (x - a)Q(x) \pmod p.$$

If also $P(b) \equiv 0 \pmod p$ for some $b \not\equiv a \pmod p$, we have

$$0 \equiv P(b) \equiv (b - a)Q(b) \pmod p.$$

Multiplying both sides of this by the inverse of $b - a$ mod p gives

$$Q(b) \equiv 0 \pmod p,$$

which similarly implies

$$Q(x) \equiv (x - b)R(x) \pmod p \quad \text{for some polynomial } R(x) \\ \text{of degree } n - 2.$$

Thus the degree of the quotient polynomial falls by 1 for each distinct solution of $P(x) \equiv 0 \pmod p$, and hence the latter congruence has at most n different solutions $\pmod p$. $\square$

Exercises

The proof of the criterion for inverses shows them connected to the gcd via the fact that $\gcd(a, n) = ma + ln$ for integers l and m. We know from Section 1.5 that an efficient way to find such integers is by the Euclidean algorithm. Thus the Euclidean algorithm is also ideal for finding inverses mod n. To find an inverse m of a, mod n, find $\gcd(a, n)$ as an explicit linear combination of a and n, and the answer $ma + nl$ contains the desired inverse m.

6.4.1. Find an inverse of 13, mod 31, by this method.

The efficiency of the Euclidean algorithm makes it feasible to find inverses for very large numbers and moduli, say, with hundreds of digits. A more difficult computational problem is finding *how many* numbers have inverses, for a given modulus n. By the criterion for inverses, the problem is to find how many of the numbers $1, 2, 3, \ldots, n - 1$ have gcd 1 with n. The number of them is denoted by $\varphi(n)$, and φ is called the *Euler phi function*. The problem of computing $\varphi(n)$ is at least as hard as recognizing whether n is prime, for the following simple reason.

6.4.2. Show that $\varphi(n) = n - 1$ if and only if n is prime.

If n is known to be prime or a prime power, then $\varphi(n)$ is easier to compute.

6.4.3. If p is prime, show that the number of multiples of p among $1, 2, 3, \ldots, p^k - 1$ is $p^{k-1} - 1$, and deduce that $\varphi(p^k) = p^{k-1}(p - 1)$.

6.4.4. Check this formula by finding $\varphi(27)$ from first principles.

Finally, if the full prime factorization of n is known, $\varphi(n)$ can be computed using the fact that $\varphi(rs) = \varphi(r)\varphi(s)$ when $\gcd(r, s) = 1$. This fact can be proved by elementary methods, but it falls more naturally out of the Chinese remainder theorem, which will be discussed in Section 6.6.

So far we have been saying *an* inverse, but in $\mathbb{Z}/n\mathbb{Z}$ we can say *the* inverse, because a number with an inverse has only one congruence class of them.

6.4.5. If $am_1 \equiv 1 \pmod{n}$ and $am_2 \equiv 1 \pmod{n}$ show that $m_1 \equiv m_2 \pmod{n}$.

Another important theorem about polynomials mod p is the "mod p binomial theorem" $(1 + x)^p \equiv 1 + x^p \pmod{p}$. First recall the ordinary binomial theorem.

6.4.6. Show by induction on n that

$$(1 + x)^n = 1 + \binom{n}{1} x + \cdots + \binom{n}{n-1} x^{n-1} + x^n,$$

where $\binom{n}{j} = \frac{n(n-1)\cdots(n-j+1)}{j(j-1)\cdots 2 \cdot 1}$ is the number of ways of choosing j things from n things.

6.4.7. Show that p divides $\binom{n}{j}$ for $1 \le j \le n - 1$, and hence conclude that $(1 + x)^p \equiv 1 + x^p \pmod{p}$.

6.5   The Theorems of Fermat and Wilson

Suppose a is any positive integer and we form the sequence of its powers: $a, a^2, a^3, \ldots$. If we reduce these powers to their values mod p, then some value must eventually repeat, because there are only p different values available. Trials with actual values of a and p suggest that the sequence of powers a^m mod p is actually periodic, and that it always includes the number 1.

For example, the sequence of powers of 2, mod 5, is

$$2, 4, 3, 1, 2, 4, 3, 1, 2, 4, 3, 1, \ldots,$$

which strongly suggests that the sequence has period 2, 4, 3, 1. Indeed it must, because the first 1 shows that $2^4 \equiv 1 \pmod{5}$, in which case $2^5 \equiv 2^1$, $2^6 \equiv 2^2$, $2^7 \equiv 2^3$, and so on, (mod 5).

It is clear from this example that the behavior of powers in arithmetic mod p depends on whether there is a power congruent to 1. Fermat's little theorem tells us that such a power always occurs. It is called Fermat's "little" theorem to distinguish it from the much more difficult "Fermat's last theorem." However, it deserves a place of its own, for both its elegance and historical importance.

Fermat's little theorem.   *If p is prime and $\gcd(a, p) = 1$, then $a^{p-1} \equiv 1 \pmod{p}$.*

Proof   The condition $\gcd(a, p) = 1$ says that a has an inverse mod p, by the criterion for inverses. The numbers $1, 2, \ldots, p - 1$ also have inverses mod p, because p is prime.

Now consider the remainders of $a, 2a, \ldots, (p-1)a$ on division by p:

$$a \bmod p, \quad 2a \bmod p, \quad \ldots, \quad (p-1)a \bmod p.$$

These remainders are the numbers $1, 2, \ldots, p-1$ again (in a different order), because they are nonzero and unequal (mod p): $ja \equiv ka$ (mod p) implies $j \equiv k$ (mod p), multiplying both sides by the inverse of a. It follows that

$$a \times 2a \times \cdots \times (p-1)a \equiv 1 \times 2 \times \cdots \times (p-1) \pmod{p},$$

that is,

$$a^{p-1} \times 1 \times 2 \times \cdots \times (p-1) \equiv 1 \times 2 \times \cdots \times (p-1) \pmod{p},$$

and therefore

$$a^{p-1} \equiv 1 \pmod{p},$$

multiplying both sides by the inverses of $1, 2, \ldots, p-1$. $\square$

Fermat's little theorem is proved by equating two different expressions for $1 \times 2 \times 3 \times \cdots \times (p-1)$. The actual value of this product, mod p, can be found by pairing factors with their inverses. The result is known as Wilson's theorem, and the following proof was given by Gauss (1801).

Wilson's theorem. *If p is prime, then*

$$1 \times 2 \times 3 \times \cdots \times (p-1) \equiv -1 \pmod{p}.$$

Proof Before pairing factors with their inverses, we have to weed out the factors that are inverse to themselves. One such factor is obviously 1, and another is -1 (which is $p-1$, mod p). To see that these are the only self-inverse factors, mod p, we note that self-inverse numbers x satisfy the quadratic equation mod p:

$$x^2 \equiv 1 \pmod{p}.$$

By Lagrange's polynomial theorem (Section 6.4) this equation has at most two solutions; hence $x \equiv 1$ (mod p) and $x \equiv -1$ (mod p) are the only ones.

Thus the factors of $1 \times 2 \times 3 \times \cdots \times (p-1)$ include exactly two that are self-inverse, 1 and $p-1$. Canceling the remaining inverse pairs leaves

$$1 \times 2 \times 3 \times \cdots \times (p - 1) \equiv 1 \times (p - 1) \equiv -1 \pmod{p},$$

as required. □

This theorem has a striking and unexpected corollary.

Wilson's primality criterion. *A natural number n is prime if and only if*

$$1 \times 2 \times 3 \times \cdots \times (n - 1) \equiv -1 \pmod{n}.$$

Proof As we have just seen, if n is prime then $1 \times 2 \times 3 \times \cdots \times (n-1) \equiv -1 \pmod{n}$.

Conversely, if n is *not* prime then the numbers $2, 3, \ldots, n - 1$ include a divisor d of n, and they also include n/d. But then $1 \times 2 \times 3 \times \cdots \times (n - 1)$ is a multiple of n and hence

$$1 \times 2 \times 3 \times \cdots \times (n - 1) \equiv 0 \not\equiv -1 \pmod{n}.$$ □

It is extremely surprising to find such a simply stated criterion for n to be prime, but unfortunately the criterion seems to have no practical value. When n is large enough for its primality to be worth asking about, it is also large enough to make $1 \times 2 \times 3 \times \cdots \times (n - 1)$ impossible to compute.

Exercises

Fermat discovered his little theorem in around 1640. As mentioned in Section 1.6, he was looking for a way to find factors of numbers of the form $2^p - 1$. His theorem can detect such factors with surprising ease, if they exist.

6.5.1. Suppose a prime $q > p$ divides $2^p - 1$, so $2^p \equiv 1 \pmod{q}$. Show the following, in turn, using Fermat's little theorem for the third step:

- If $2^a \equiv 1 \pmod{q}$ and $2^b \equiv 1 \pmod{q}$ with $a > b$ then $2^{a-b} \equiv 1 \pmod{q}$.

- If $2^a \equiv 1 \pmod{q}$ and $2^b \equiv 1 \pmod{q}$ then $2^{\gcd(a,b)} \equiv 1 \pmod{q}$.

- If $2^p \equiv 1 \pmod{q}$ for a prime p, then p divides $q - 1$.

- q is of the form $kp + 1$ for some integer k.

The first number on Fermat's hit list was $2^{37} - 1$. According to Exercise 6.5.1, any prime divisor > 37 must be one of $37 + 1$, $2 \times 37 + 1$, $3 \times 37 + 1$, ... The first prime in this sequence is $6 \times 31 + 1 = 223$ and ... bingo!

6.5.2. Check that 223 divides $2^{37} - 1$.

If the prime q divides $2^n - 1$ and n is *not* prime, then n does not necessarily divide $q - 1$.

6.5.3. Find an n such that 31 divides $2^n - 1$ but n does not divide 30.

However, if m is the *least* positive exponent such that the prime q divides $2^m - 1$ it is true that m divides $q - 1$.

6.5.4. If m is the least positive exponent such that $2^m \equiv 1 \pmod{q}$, show that m divides any positive n such that $2^n \equiv 1 \pmod{q}$ (in particular, m divides the exponent $q - 1$ given by Fermat's little theorem).

This fact greatly shortens the search for divisors of the Fermat number $2^{2^5} + 1 = 2^{32} + 1$. Any prime divisor q of $2^{32} + 1$ also divides $(2^{32} + 1)(2^{32} - 1) = 2^{64} - 1$, and 64 is the least m such that q divides $2^m - 1$, for the following reason.

6.5.5. Show that the least positive m such that $2^m \equiv 1 \pmod{q}$ is a divisor of 64. Conclude, using the fact that $2^{32} \equiv -1 \pmod{q}$, that $m = 64$.

6.5.6. Deduce from Exercise 6.5.5 that any prime divisor of $2^{2^5} + 1$ is of the form $64k + 1$.

If Fermat had followed his own train of thought this far he would not have made the mistake of thinking that all the numbers $2^{2^h} + 1$ are prime. In fact, this is precisely how Euler discovered the divisor 641 of $2^{2^5} + 1$.

Wilson's theorem was first published without proof, in a book by Edward Waring in 1770. The first proof was given by Lagrange in 1771, and he also used it to find the primes p for which -1 is a square, mod p.

6.5.7. If $p = 2$, then -1 is certainly a square mod p. Why?

6.5.8. If $p = 4n + 3$, use congruences mod 4 to show that -1 is not a square mod p.

The most challenging case is where $p = 4n+1$, in which case Wilson's theorem is helpful, combined with the fact that $1 \times 2 \times \cdots \times (p-1) = 1 \times 2 \times \cdots \times 4n$.

6.5.9. Show that $1 \times 2 \times \cdots \times 4n \equiv (1 \times 2 \times \cdots \times 2n)^2 \pmod{4n+1}$, and hence conclude from Wilson's theorem that $-1 \equiv (1 \times 2 \times \cdots \times 2n)^2 \pmod{p}$ when $p = 4n + 1$ is prime.

6.6   The Chinese Remainder Theorem

The behavior of numbers mod n is quite complicated when n is not prime. As we have seen, there are nonzero numbers without inverses, and finding all the numbers with inverses is tied up with the hard problem of computing $\varphi(n)$. Some relief from this situation is obtained by "factorizing" the ring $\mathbb{Z}/n\mathbb{Z}$ into smaller and simpler rings. The germ of this idea was discovered by Chinese mathematicians around 300 A.D., and various generalizations of it are now called the *Chinese remainder theorem*.

The theorem grows out of the discovery that a number can be known modulo lm if it is known modulo l and m. For example, the number 25 is completely determined, mod 77, by the two remainders $25 \bmod 7 = 4$ and $25 \bmod 11 = 3$. The reason no other number < 77 gives these remainders is that *all 77 pairs of remainders occur*, so there is exactly one pair for each of the numbers $0, 1, 2, 3, \ldots, 76$. This is proved by an algorithm that actually obtains the natural number x with a given pair of remainders. To do this, the Chinese used what they called the *method of finding 1*.

The method (in its basic form) assumes we have a relatively prime pair l and m, so that $\gcd(l, m) = 1$, and the Euclidean algorithm can be used to express 1 as a linear combination of l and m.

Example.   To obtain x with $x \bmod 7 = 4$ and $x \bmod 11 = 2$.

- First express $1 = \gcd(11, 7)$ as a linear combination of 11 and 7, say,

$$1 = 2 \times 11 - 3 \times 7.$$

- Then express 4 and 2 as multiples of this combination:

$$4 = 8 \times 11 - 12 \times 7 \quad \text{and} \quad 2 = 4 \times 11 - 6 \times 7.$$

- This gives 4 and 2 as remainders on division by 7 and 11, respectively,

$$4 = 8 \times 11 \bmod 7 \quad \text{and} \quad 2 = -6 \times 7 \bmod 11.$$

- And the *sum* of these multiples of 11 and 7 has the same remainders,

$$4 = (8 \times 11 - 6 \times 7) \bmod 7 \quad \text{and} \quad 2 = (8 \times 11 - 6 \times 7) \bmod 11.$$

- Hence the solution is $x = (8 \times 11 - 6 \times 7) = 88 - 42 = 46$.

The traditional Chinese remainder theorem is about *determining* a number by a pair (or triple, quadruple, etc.) of smaller numbers. However, we can also add and multiply numbers by adding and multiplying the corresponding pairs. We want the sum of the pairs for x_1 and x_2 to be the pair for $x_1 + x_2$, so the rule for adding pairs is

$$(x_1 \bmod l, x_1 \bmod m) + (x_2 \bmod l, x_2 \bmod m)$$
$$= (x_1 + x_2 \bmod l, x_1 + x_2 \bmod m),$$

and similarly the rule for multiplying pairs is

$$(x_1 \bmod l, x_1 \bmod m)(x_2 \bmod l, x_2 \bmod m)$$
$$= (x_1 x_2 \bmod l, x_1 x_2 \bmod m).$$

The fully fledged Chinese remainder theorem includes this arithmetic of pairs by describing the ring $\mathbb{Z}/lm\mathbb{Z}$ as a "product" of the rings $\mathbb{Z}/l\mathbb{Z}$ and $\mathbb{Z}/m\mathbb{Z}$. It is called the *direct product* $\mathbb{Z}/l\mathbb{Z} \times \mathbb{Z}/m\mathbb{Z}$ and its members are pairs of congruence classes $(x + l\mathbb{Z}, x + m\mathbb{Z})$, added and multiplied according to the rules

$$(x_1 + l\mathbb{Z}, x_1 + m\mathbb{Z}) + (x_2 + l\mathbb{Z}, x_2 + m\mathbb{Z}) = (x_1 + x_2 + l\mathbb{Z}, x_1 + x_2 + m\mathbb{Z}),$$
$$(x_1 + l\mathbb{Z}, x_1 + m\mathbb{Z})(x_2 + l\mathbb{Z}, x_2 + m\mathbb{Z}) = (x_1 x_2 + l\mathbb{Z}, x_1 x_2 + m\mathbb{Z}).$$

These rules are just a translation, into the language of congruence classes, of the rules just stated for pairs of remainders mod l and mod m.

$\mathbb{Z}/l\mathbb{Z} \times \mathbb{Z}/m\mathbb{Z}$ is not strictly identical with $\mathbb{Z}/lm\mathbb{Z}$, because its members are pairs rather than single congruence classes, but it *be-*

haves the same in a sense that will be explained in the proof of the theorem. We say that $\mathbb{Z}/l\mathbb{Z} \times \mathbb{Z}/m\mathbb{Z}$ is *isomorphic* to $\mathbb{Z}/lm\mathbb{Z}$ (from the Greek for "same form") and write

$$\mathbb{Z}/lm\mathbb{Z} \cong \mathbb{Z}/l\mathbb{Z} \times \mathbb{Z}/m\mathbb{Z}.$$

In practice, there is no harm in saying that the ring $\mathbb{Z}/lm\mathbb{Z}$ is the direct product $\mathbb{Z}/l\mathbb{Z} \times \mathbb{Z}/m\mathbb{Z}$.

Chinese remainder theorem.   *If* $\gcd(l, m) = 1$ *then*

$$\mathbb{Z}/lm\mathbb{Z} \cong \mathbb{Z}/l\mathbb{Z} \times \mathbb{Z}/m\mathbb{Z}.$$

Proof   We begin by letting the congruence class $x + lm\mathbb{Z}$ correspond to the pair $(x+l\mathbb{Z}, x+m\mathbb{Z})$. Because there are lm classes $x+lm\mathbb{Z}$, also l classes $a+l\mathbb{Z}$ and m classes $b+m\mathbb{Z}$, the latter form lm pairs. Thus the correspondence will be one-to-one between $\mathbb{Z}/lm\mathbb{Z}$ and $\mathbb{Z}/l\mathbb{Z} \times \mathbb{Z}/m\mathbb{Z}$ provided each pair $(a + l\mathbb{Z}, b + m\mathbb{Z})$ is $(x + l\mathbb{Z}, x + m\mathbb{Z})$ for some x.

It suffices to find integers u and v with $um \bmod l = a$ and $vl \bmod m = b$, because $x = um + vl$ will then give $x \bmod l = a$ and $x \bmod m = b$, as required. This is where we use the fact that $\gcd(l, m) = 1$. By Section 1.5, and the "method of finding 1,"

$$\gcd(l, m) = 1 \Rightarrow 1 = rl + sm \text{ for some integers } r \text{ and } s,$$
$$\Rightarrow a = arl + asm \quad \text{and} \quad b = brl + bsm,$$
$$\Rightarrow asm = a - arl \quad \text{and} \quad brl = b - bsm,$$
$$\Rightarrow asm \bmod l = a, \quad \text{and} \quad brl \bmod m = b.$$

Hence suitable integers are $u = as$ and $v = br$, and $x = asm + brl$.

By definition of the sum of pairs, the pair in $\mathbb{Z}/l\mathbb{Z} \times \mathbb{Z}/m\mathbb{Z}$ corresponding to a sum in $\mathbb{Z}/lm\mathbb{Z}$ is the sum of the corresponding pairs. Products similarly correspond to products, so we have a one-to-one correspondence between $\mathbb{Z}/lm\mathbb{Z}$ and $\mathbb{Z}/l\mathbb{Z} \times \mathbb{Z}/m\mathbb{Z}$ that preserves sums and products. This is precisely what we mean by $\mathbb{Z}/lm\mathbb{Z} \cong \mathbb{Z}/l\mathbb{Z} \times \mathbb{Z}/m\mathbb{Z}$.   □

Exercises

It is helpful to work out the actual pairs for a small case of $\mathbb{Z}/lm\mathbb{Z}$, for example:

6.6.1. For each x in $\mathbb{Z}/15\mathbb{Z}$, work out the corresponding pair $(x \bmod 3,$ $x \bmod 5)$ in $\mathbb{Z}/3\mathbb{Z} \times \mathbb{Z}/5\mathbb{Z}$. What do you notice about the pairs for invertible x?

The isomorphism between $\mathbb{Z}/l\mathbb{Z} \times \mathbb{Z}/m\mathbb{Z}$ and $\mathbb{Z}/lm\mathbb{Z}$ looks like a formality once the one-to-one correspondence has been discovered, but the structure it gives to the set of pairs $(a+l\mathbb{Z}, b+m\mathbb{Z})$ is surprisingly helpful. For example, because classes $x+lm\mathbb{Z}$ behave the same as the corresponding pairs $(a+l\mathbb{Z}, b+m\mathbb{Z})$, it follows in particular that classes with inverses correspond to pairs with inverses.

6.6.2. Show that a pair $(a + l\mathbb{Z}, b + m\mathbb{Z})$ with an inverse corresponds in turn to an $a + l\mathbb{Z}$ with an inverse and a $b + m\mathbb{Z}$ with an inverse.

Now, by definition of the Euler φ function, there are $\varphi(l)$ such classes $a+l\mathbb{Z}$, and $\varphi(m)$ such classes $b+m\mathbb{Z}$. This gives the *multiplicative property* of φ.

6.6.3. Deduce that there are $\varphi(l)\varphi(m)$ invertible pairs $(a+l\mathbb{Z}, b+m\mathbb{Z})$ and hence conclude: if $\gcd(l, m) = 1$ then $\varphi(lm) = \varphi(l)\varphi(m)$.

As mentioned in the exercises to Section 6.4, this property of the Euler φ function enables us to compute $\varphi(n)$ when the prime factorization of n is known.

6.6.4. Show that if $n = p_1^{e_1} p_2^{e_2} \cdots p_k^{e_k}$ is the prime factorization of n then

$$\varphi(n) = p_1^{e_1-1}(p_1 - 1)p_2^{e_2-1}(p_2 - 1) \cdots p_k^{e_k-1}(p_k - 1).$$

6.7   Squares mod p

In arithmetic mod p the analogs of linear and quadratic equations are linear and quadratic congruences, and they are solved in an analogous way—up to a point. Because the ordinary operations of arithmetic are valid mod p, we can solve the linear congruence

$$ax + b \equiv 0 \pmod{p}$$

by subtracting b from both sides, then multiplying both sides by the mod p inverse of a. If we write this inverse as $1/a$, then the solution looks the same as in ordinary algebra: $x = -b/a$. The difference, of course, is that the inverse mod p of a is the (congruence class of the) solution m of $ma + np = 1$, which we find by applying the Euclidean algorithm to express $1 = \gcd(a, p)$ in the form $ma + np$.

Likewise, the quadratic congruence $ax^2 + bx + c \equiv 0 \pmod{p}$ can be solved, as in ordinary algebra, by "completing the square." We find

$$ax^2 + bx + c \equiv 0 \pmod{p}$$
$$\Rightarrow a\left(x^2 + \frac{b}{a}x\right) + c \equiv 0 \pmod{p}$$
$$\Rightarrow a\left(x^2 + \frac{b}{a}x + \frac{b^2}{4a^2}\right) + c - \frac{b^2}{4a} \equiv 0 \pmod{p}$$
$$\Rightarrow a\left(x + \frac{b}{2a}\right)^2 \equiv \frac{b^2}{4a} - c \pmod{p}$$
$$\Rightarrow \left(x + \frac{b}{2a}\right)^2 \equiv \frac{b^2 - 4ac}{(2a)^2} \pmod{p}$$

by various applications of $+$, $-$, $\times$, and $\div$ mod p. The big difference is in the next step: finding the "square root" mod p, and indeed deciding whether it exists. This turns out to be a deep and interesting problem, to which we shall devote the next few sections of this chapter. It so happens that exactly half the numbers $1, 2, 3, \ldots, p-1$ are squares mod p, but the rule for finding them is quite mysterious and unexpected.

The first step toward finding which numbers are squares mod p is fairly simple, thanks to Lagrange's polynomial theorem (Section 6.4). We can confine attention to odd primes p, because the only numbers mod 2 are 0 and 1, and these are obviously squares for any modulus.

Euler's criterion. *For an odd prime p, $a \not\equiv 0$ is a square* mod p $\Leftrightarrow a^{\frac{p-1}{2}} \equiv 1$ mod p.

Proof The ($\Rightarrow$) direction is an easy consequence of Fermat's little theorem (Section 6.5):

$$a \text{ is a square mod } p \Rightarrow a \equiv b^2 \pmod{p} \text{ for some } b$$
$$\Rightarrow a^{\frac{p-1}{2}} \equiv b^{p-1} \equiv 1 \pmod{p}$$
$$\text{by Fermat's little theorem}$$

To prove the ($\Leftarrow$) direction we first observe that exactly half of the numbers $1, 2, 3, \ldots, p - 1$ are squares mod p because:

- No two of $1^2, 2^2, 3^2, \ldots, \left(\frac{p-1}{2}\right)^2$ are congruent mod p. This is because $i^2 \equiv j^2 \pmod{p}$ implies $(i - j)(i + j) \equiv 0 \pmod{p}$, which is impossible for distinct i and j among $1, 2, 3, \ldots, \frac{p-1}{2}$, because $i \pm j \not\equiv 0 \pmod{p}$.
- $(p - k)^2 \equiv (-k)^2 \equiv k^2 \pmod{p}$. Hence the only values squares can take are the $\frac{p-1}{2}$ distinct values $1^2, 2^2, 3^2, \ldots, \left(\frac{p-1}{2}\right)^2$.

Thus there are $\frac{p-1}{2}$ nonzero squares mod p. By the first part of the proof they are all solutions of $x^{\frac{p-1}{2}} \equiv 1 \pmod{p}$, and by Lagrange's polynomial theorem there are no other solutions of this congruence. Hence, if a is not a square mod p then $a^{\frac{p-1}{2}} \not\equiv 1 \pmod{p}$. $\square$

Squares mod p are often called *quadratic residues* mod p, and nonsquares are called quadratic *nonresidues*. The terminology is borrowed from Latin, where the same word means both "square" and "quadratic," and it seems misleading to use it when "squares mod p" and "nonsquares mod p" are available. A useful notation for saying whether or not a nonzero a is a square mod p is the *Legendre symbol*, $\left(\frac{a}{p}\right)$. This symbol is also called the *quadratic character* of $a \pmod{p}$, and is defined by

$$\left(\frac{a}{p}\right) = \begin{cases} 1 & \text{if } a \text{ is a square mod } p \\ -1 & \text{if } a \text{ is a nonsquare mod } p \end{cases}$$

The value of -1 for nonsquares actually comes out the proof of Euler's criterion, if one looks closely, leading to the following.

Restatement of Euler's criterion. $\left(\frac{a}{p}\right) \equiv a^{\frac{p-1}{2}} \pmod{p}$.

Proof $\left(a^{\frac{p-1}{2}}\right)^2 \equiv a^{p-1} \equiv 1 \pmod{p}$ by Fermat's little theorem, and $x^2 \equiv 1 \pmod{p}$ has only the two solutions $x = 1$ and $x = -1$ by Lagrange's polynomial theorem. Therefore, the only possible values $\pmod{p}$ of $a^{\frac{p-1}{2}}$ are 1, which it takes for squares a, and -1, which it necessarily takes for nonsquares a.

Thus $a^{\frac{p-1}{2}} \equiv \left(\frac{a}{p}\right) \pmod{p}$, by definition of the Legendre symbol. $\square$

Exercises

There is another proof of Euler's criterion, which is shorter and more enlightening, but dependent on a harder theorem: the existence of primitive roots. A *primitive root* mod p is a number r such that each of $1, 2, 3, \ldots, p-1$ is congruent to a power of r, mod p.

6.7.1. Show that 2 is a primitive root mod 5, but not a primitive root mod 7. Find a primitive root mod 7.

The existence of a primitive root for each prime p was conjectured by Euler and proved by Gauss (1801). All proofs I am aware of use Lagrange's polynomial theorem plus some extra ingenuity, so the existence of primitive roots should probably be regarded as a harder theorem than Euler's criterion. However, it also throws more light on Euler's criterion.

6.7.2. If r is a primitive root mod p, show that the nonzero squares mod p are the even powers of r. Deduce that there are $\frac{p-1}{2}$ nonzero squares mod p.

6.7.3. Deduce the ($\Leftarrow$) direction of Euler's criterion from Exercise 6.7.2.

The existence of primitive roots can also be used to prove analogous theorems about cubes mod p, and so on. These results are not as complete as Euler's criterion for squares, because they depend on p. Here is what we can say about cubes.

6.7.4. If 3 divides $p - 1$ and r is a primitive root mod p, show that the nonzero cubes mod p are $1, r^3, r^6, \ldots$. Deduce that a is a cube mod $p \Leftrightarrow a^{\frac{p-1}{3}} \equiv 1 \pmod{p}$.

6.7.5. If 3 does not divide $p - 1$, which numbers are cubes mod p?

6.8*　The Quadratic Character of -1 and 2

Euler's criterion does not immediately tell us which a are squares modulo a given odd prime p or the moduli p for which a given a is a square. However, it can be used to obtain this information explicitly for the two important values $a = -1$ and $a = 2$.

Quadratic character of −1. *For any odd prime p, −1 is a square mod p ⇔ p = 4n + 1 for some integer n.*

Proof By Euler's criterion,

$$-1 \text{ is a square mod } p \Leftrightarrow (-1)^{\frac{p-1}{2}} \equiv 1 \pmod{p}$$

$$\Leftrightarrow \frac{p-1}{2} \text{ is even}$$

$$\Leftrightarrow p = 4n+1 \text{ for some integer } n. \qquad \square$$

To find the quadratic character of 2 we have the harder job of evaluating $2^{\frac{p-1}{2}}$ mod p. This can be done by manipulating the product $1 \times 2 \times 3 \times \cdots \times (p-1) \pmod{p}$ into the form

$$2^{\frac{p-1}{2}}(-1)^{\frac{p-1}{4}} \times 1 \times 2 \times 3 \times \cdots \times (p-1) \quad \text{if } \frac{p-1}{2} \text{ is even,}$$

$$2^{\frac{p-1}{2}}(-1)^{\frac{p+1}{4}} \times 1 \times 2 \times 3 \times \cdots \times (p-1) \quad \text{if } \frac{p-1}{2} \text{ is odd.}$$

From this we conclude (by canceling $1, 2, 3, \ldots, p-1$) that

$$2^{\frac{p-1}{2}} \equiv \begin{cases} (-1)^{\frac{p-1}{4}} \pmod{p} & \text{if } \frac{p-1}{2} \text{ is even} \\ (-1)^{\frac{p+1}{4}} \pmod{p} & \text{if } \frac{p-1}{2} \text{ is odd.} \end{cases}$$

The manipulation becomes clearer with an accompanying example, say $p = 11$.

In the product,

$$1 \times 2 \times 3 \times 4 \times 5 \times 6 \times 7 \times 8 \times 9 \times 10,$$

separate the even and odd factors,

$$(2 \times 4 \times 6 \times 8 \times 10) \times 1 \times 3 \times 5 \times 7 \times 9.$$

Extract 2 from the $(p-1)/2$ even factors,

$$2^5(1 \times 2 \times 3 \times 4 \times 5) \times 1 \times 3 \times 5 \times 7 \times 9$$

so that even factors $> (p-1)/2$ are lost and odd factors $\leq (p-1)/2$ are repeated.

$$2^5(1 \times 2 \times 3 \times 4 \times 5) \times \underline{1} \times \underline{3} \times \underline{5} \times 7 \times 9$$

Give the repeated factors − signs, inserting factors of −1 to compensate,

$$2^5(1 \times 2 \times 3 \times 4 \times 5) \times (-1)^3(-1) \times (-3) \times (-5) \times 7 \times 9.$$

Replace each odd factor $-n$ by $p - n$, which is even and $> (p - 1)/2$,

$$2^5(1 \times 2 \times 3 \times 4 \times 5) \times (-1)^3 \underline{10} \times \underline{8} \times \underline{6} \times 7 \times 9$$

so that the new product $\equiv$ the old (mod p), and includes all of $1, 2, 3, \ldots, p - 1$.

It is clear from this example why the exponent of 2 is $(p - 1)/2$, because this is the number of even numbers among $1, 2, 3, \ldots, p - 1$. The exponent of -1 is the number of odd numbers $\leq (p - 1)/2$, namely, $(p - 1)/4$ if $(p - 1)/2$ is even, and $(p + 1)/4$ if $(p - 1)/2$ is odd, hence the result is as claimed.

From the value of $2^{\frac{p-1}{2}}$ mod p we can now deduce an explicit description of the odd prime moduli for which 2 is a square.

Quadratic character of 2.   *For any odd prime p, 2 is a square mod $p \Leftrightarrow p = 8n \pm 1$ for some integer n.*

Proof   By Euler's criterion, 2 is a square mod $p \Leftrightarrow 2^{\frac{p-1}{2}} \equiv 1 \pmod{p}$, so it suffices to evaluate $2^{\frac{p-1}{2}}$ (using the expression $(-1)^{\frac{p-1}{4}}$ for $\frac{p-1}{2}$ even, and $(-1)^{\frac{p+1}{4}}$ for $\frac{p-1}{2}$ odd) for the possible odd values of p. Apart from $8n \pm 1$, the other odd values are $8n \pm 3$, and we find

$$p = 8n + 1 \Rightarrow \frac{p - 1}{2} \text{ even} \Rightarrow 2^{\frac{p-1}{2}} \equiv (-1)^{\frac{p-1}{4}} \equiv (-1)^{\frac{8n}{4}} \equiv 1 \pmod{p}$$

$$p = 8n - 1 \Rightarrow \frac{p - 1}{2} \text{ odd} \Rightarrow 2^{\frac{p-1}{2}} \equiv (-1)^{\frac{p+1}{4}} \equiv (-1)^{\frac{8n}{4}} \equiv 1 \pmod{p}$$

$$p = 8n + 3 \Rightarrow \frac{p - 1}{2} \text{ odd} \Rightarrow 2^{\frac{p-1}{2}} \equiv (-1)^{\frac{p+1}{4}} \equiv (-1)^{\frac{8n+4}{4}}$$
$$\equiv -1 \pmod{p}$$

$$p = 8n - 3 \Rightarrow \frac{p - 1}{2} \text{ even} \Rightarrow 2^{\frac{p-1}{2}} \equiv (-1)^{\frac{p-1}{4}} \equiv (-1)^{\frac{8n-4}{4}}$$
$$\equiv -1 \pmod{p}$$

as required.   □

The calculation of $2^{\frac{p-1}{2}}$ mod p may seem like a lucky accident, but there is reason to believe in advance that it will work. By Wilson's theorem, $1 \times 2 \times 3 \times \cdots \times (p - 1) \equiv -1 \pmod{p}$, and by Euler's criterion $2^{\frac{p-1}{2}} \equiv \pm 1 \pmod{p}$. Therefore, if we can extract the factor $2^{\frac{p-1}{2}}$ from $1 \times 2 \times 3 \times \cdots \times (p - 1)$ (which we obviously can, from the even numbers), then the remaining factor must be $\equiv \pm 1 \pmod{p}$.

Exercises

The description of the quadratic character of 2 can be condensed as follows.

6.8.1. Show that $\left(\frac{2}{p}\right) = (-1)^{\frac{p^2-1}{8}}$.

As suggested earlier, the calculation of $\left(\frac{2}{p}\right)$ from $1 \times 2 \times 3 \times \cdots \times (p-1)$ can be expected to work, so it is mainly a matter of shuffling the factors until we get what we want. A more imaginative calculation of $\left(\frac{2}{p}\right)$, using $i = \sqrt{-1}$ and de Moivre's formula, is given in Scharlau and Opolka (1985). The main steps follow.

6.8.2.* Using the fact that $2 = \frac{(1+i)^2}{i}$ and Euler's criterion, show that

$$\left(\frac{2}{p}\right) \equiv \frac{(1+i)^p}{i^{\frac{p-1}{2}}(1+i)} \equiv \frac{1+i^p}{i^{\frac{p-1}{2}}(1+i)} \pmod{p}.$$

6.8.3.* Using the fact that $i = \cos\frac{\pi}{2} + i\sin\frac{\pi}{2}$, show that

$$\frac{1+i^p}{i^{\frac{p-1}{2}}(1+i)} = \frac{(1+i^p)i^{-p/2}}{(1+i)i^{-1/2}} = \frac{\cos(p\pi/4)}{\cos(\pi/4)}.$$

6.8.4.* Deduce from Exercises 6.8.2* and 6.8.3* that

$$\left(\frac{2}{p}\right) = \begin{cases} 1 & \text{if } p \equiv \pm 1 \pmod 8 \\ -1 & \text{if } p \equiv \pm 3 \pmod 8 \end{cases} = (-1)^{\frac{p^2-1}{8}}.$$

6.9* Quadratic Reciprocity

The Euler criterion may be used to find $\left(\frac{q}{p}\right)$ for various fixed primes q, but it is hard to see any general pattern to the results. Legendre discovered the secret: *knowing whether q is a square mod p depends on knowing whether p is a square mod q*. The exact relationship between the primes p and q is expressed by the *law of quadratic reciprocity*, the fundamental theorem about squares modulo odd primes, first proved by Gauss (1801):

> *For odd primes p and q,*
> > *if p and q are both of the form 4n + 3 then*
> > > *p is a square mod q ⇔ q is not a square mod p,*

otherwise

p *is a square mod* q $\Leftrightarrow$ q *is a square mod* p.

The law is usually presented more concisely with the help of the Legendre symbol. When p and q are both of the form $4n+3$, quadratic reciprocity says that $\left(\frac{p}{q}\right)$ and $\left(\frac{q}{p}\right)$ have opposite signs, and hence their product is -1. Otherwise, it says that $\left(\frac{p}{q}\right)$ and $\left(\frac{q}{p}\right)$ have the same sign and hence their product is 1. All this is captured by the single equation

$$\left(\frac{p}{q}\right)\left(\frac{q}{p}\right) = (-1)^{\frac{p-1}{2}\frac{q-1}{2}}.$$

The law of quadratic reciprocity has been proved more often than any other theorem in mathematics except the Pythagorean theorem. However, it is a more difficult theorem, and none of its proofs is completely transparent. One of the shortest was given by George Rousseau (1991). It produces the result like a rabbit out of a hat, but at least the trick can be done with readily available materials: Wilson's theorem and Euler's criterion. Rousseau's proof may be compared with the computation of $\left(\frac{2}{p}\right)$ in Section 6.8*. It is a manipulation of certain products, mod p and mod q, but this time with the Chinese remainder theorem playing a crucial role. To simplify formulas, we use the standard abbreviation $n!$ for $1 \times 2 \times 3 \times \cdots \times n$.

Quadratic reciprocity. *For any odd primes p and q,*

$$\left(\frac{p}{q}\right)\left(\frac{q}{p}\right) = (-1)^{\frac{p-1}{2}\frac{q-1}{2}}.$$

Proof Consider the (congruence classes of the) invertible numbers mod pq. By the Chinese remainder theorem, each such number x can be faithfully represented by the pair $(x \bmod p, x \bmod q)$. When we multiply such pairs, the first components are multiplied mod p, and the second components are multiplied mod q.

We want to form the product of all such pairs for the invertible x between 1 and $(pq-1)/2$ inclusive. As we know from Section 6.4, the invertible x are those that are multiples of neither p nor q. We form their product mod p by multiplying the nonmultiples of p, then dividing by the multiples of q. The nonmultiples of p form the sequence

$$1, 2, \ldots, p - 1; p + 1, p + 2, \ldots 2p - 1; \ldots .$$

Taking these mod p, we get $(q - 1)/2$ sequences $1, 2, \ldots, p - 1$, followed by the "half sequence" $1, 2, \ldots, (p-1)/2$. By Wilson's theorem, the mod p product of $1, 2, \ldots, p - 1$ is -1, hence the mod p product of all nonmultiples of p between 1 and $(pq - 1)/2$ is

$$(-1)^{\frac{q-1}{2}}((p - 1)/2)!.$$

Now we divide this by the multiples $q, 2q, \ldots, ((p - 1)/2)q$ of q between 1 and $(pq - 1)/2$. Their product is

$$q^{\frac{p-1}{2}}((p - 1)/2)!,$$

so division gives $(-1)^{\frac{q-1}{2}}/q^{\frac{p-1}{2}}$. By Euler's criterion, $q^{\frac{p-1}{2}} \equiv \left(\frac{q}{p}\right)$ (mod p), which is either 1 or -1, so it makes no difference whether we multiply or divide by it: the mod p product of the invertible x from 1 to $(pq - 1)/2$ is $\left(\frac{q}{p}\right)(-1)^{\frac{q-1}{2}}$.

Similarly, the mod q product of the invertible x is $\left(\frac{p}{q}\right)(-1)^{\frac{p-1}{2}}$. Hence the product of the pairs $(x \bmod p, x \bmod q)$ for invertible x from 1 to $(pq - 1)/2$ is

$$\left(\left(\frac{q}{p}\right)(-1)^{\frac{q-1}{2}}, \left(\frac{p}{q}\right)(-1)^{\frac{p-1}{2}}\right). \tag{1}$$

Now we compute the same product in a second way, which allows it to be expressed without Legendre symbols. Equating the two expressions for the product will give a relation between $\left(\frac{q}{p}\right)$ and $\left(\frac{p}{q}\right)$.

The Chinese remainder theorem says that the pairs $(x \bmod p, x \bmod q)$ for invertible x from 1 to $pq - 1$ are the (a, b) with $1 \leq a \leq p - 1$ and $1 \leq b \leq q - 1$. Also, the pair $(pq - x \bmod p, pq - x \bmod q)$, that is, $(-x \bmod p, -x \bmod q)$, equals $(-a, -b)$ if $(x \bmod p, x \bmod q)$ equals (a, b). It follows that the pairs $(x \bmod p, x \bmod q)$ for $1 \leq x \leq (pq - 1)/2$ include (a, b) if and only if they do *not* include $(-a, -b)$.

Thus the product of the $(x \bmod p, x \bmod q)$, for the invertible x from 1 to $(pq - 1)/2$, is the product (up to a $\pm$ sign) of any set of pairs that includes (a, b) if and only if it does not include $(-a, -b)$. One such set is

$$(a, b) \qquad \text{for} \quad 1 \le a \le p - 1, \quad 1 \le b \le (q - 1)/2.$$

And the product of the members of this set is

$$\pm \left((p - 1)!^{\frac{q-1}{2}}, ((q - 1)/2)!^{p-1} \right),$$

because each value of a, $1 \le a \le p - 1$, occurs in $(q - 1)/2$ pairs, and each value of b, $1 \le b \le (q - 1)/2$, occurs in $p - 1$ pairs.

Bearing in mind that the first component is taken mod p, Wilson's theorem gives $(p - 1)! \equiv -1 \pmod{p}$, hence the first component $\equiv (-1)^{\frac{q-1}{2}} \pmod{p}$.

The second component is taken mod q, so Wilson's theorem gives

$$-1 \equiv (q - 1)! \pmod{q}$$
$$\equiv 1 \times 2 \times \cdots \times ((q - 1)/2)$$
$$\times (-((q - 1)/2) \times \cdots \times (-2) \times (-1) \pmod{q}$$
$$\equiv ((q - 1)/2)!^2 (-1)^{\frac{q-1}{2}} \pmod{q}.$$

Therefore

$$((q - 1)/2)!^2 \equiv (-1)(-1)^{\frac{q-1}{2}} \pmod{q},$$

and hence, raising both sides to the power $\frac{p-1}{2}$, we get the second component

$$((q - 1)/2)!^{p-1} \equiv (-1)^{\frac{p-1}{2}} (-1)^{\frac{p-1}{2}\frac{q-1}{2}} \pmod{q}.$$

Thus the second expression for the product simplifies to

$$\pm \left((-1)^{\frac{q-1}{2}}, (-1)^{\frac{p-1}{2}} (-1)^{\frac{p-1}{2}\frac{q-1}{2}} \right). \tag{2}$$

Equating (1) and (2) we get either

$$\left(\frac{q}{p} \right) = 1 \quad \text{and} \quad \left(\frac{p}{q} \right) = (-1)^{\frac{p-1}{2}\frac{q-1}{2}}$$

or

$$\left(\frac{q}{p} \right) = -1 \quad \text{and} \quad \left(\frac{p}{q} \right) = -(-1)^{\frac{p-1}{2}\frac{q-1}{2}}.$$

In either case, the product of the two equations is

$$\left(\frac{p}{q} \right) \left(\frac{q}{p} \right) = (-1)^{\frac{p-1}{2}\frac{q-1}{2}}.$$

$\square$

Exercises

Once we know the primes p that are squares modulo an odd prime q we can recognize *all* the squares mod q.

6.9.1. Show that if $P = p_1 p_2 \cdots p_k$ is the prime factorization of P then

$$P \text{ is a square mod } q \quad \Leftrightarrow \quad \left(\frac{p_1}{q} \right) \left(\frac{p_2}{q} \right) \cdots \left(\frac{p_k}{q} \right) = 1.$$

This result suggests a natural extension of the Legendre symbol to all numbers $P \not\equiv 0 \pmod{q}$: if $P = p_1 p_2 \cdots p_k$ is the prime factorization of P, let

$$\left(\frac{P}{q} \right) = \left(\frac{p_1}{q} \right) \left(\frac{p_2}{q} \right) \cdots \left(\frac{p_k}{q} \right).$$

6.9.2. Deduce from this definition that the Legendre symbol is *multiplicative*:

$$\left(\frac{PQ}{q} \right) = \left(\frac{P}{q} \right) \left(\frac{Q}{q} \right) \quad \text{for any } P, Q \not\equiv 0 \pmod{q}.$$

Also, of course, $\left(\frac{P}{q} \right)$ depends only on the congruence class of P, mod q, so we can replace P by its remainder on division by q. Using this fact, the multiplicative property, and quadratic reciprocity, the computation of Legendre symbols is greatly simplified.

6.9.3. Justify each step in the following computations.

$$\left(\frac{5}{31} \right) = \left(\frac{31}{5} \right) = \left(\frac{1}{5} \right) = 1$$

$$\left(\frac{7}{31} \right) = -\left(\frac{31}{7} \right) = -\left(\frac{3}{7} \right) = \left(\frac{7}{3} \right) = \left(\frac{1}{3} \right) = 1$$

$$\left(\frac{11}{31} \right) = -\left(\frac{31}{11} \right) = -\left(\frac{9}{11} \right) = -\left(\frac{3}{11} \right)^2 = -1$$

$$\left(\frac{13}{31} \right) = \left(\frac{31}{13} \right) = \left(\frac{5}{13} \right) = \left(\frac{13}{5} \right) = \left(\frac{3}{5} \right) = \left(\frac{5}{3} \right) = \left(\frac{2}{3} \right) = -1.$$

6.9.4. Use similar steps to show that $\left(\frac{19}{31} \right) = 1$. What is 19 the square of, mod 31?

To complete our toolkit for recognizing whether P is a square mod q we need a rule for evaluating $\left(\frac{2}{q} \right)$, because any $P > 1$ is a product of odd primes and 2s. This why we worked out $\left(\frac{2}{q} \right)$ in Section 6.8*. (When

$\left(\frac{2}{3}\right)$ came up in Exercise 6.9.3, we were able to evaluate it by inspection, simply because there are so few squares mod 3.) The formula we found,

$$\left(\frac{2}{p}\right) = (-1)^{\frac{p^2-1}{8}},$$

is known as a *supplement* to the law of quadratic reciprocity.

6.10　Discussion

Congruences and Congruence Classes

Gauss (1801) used his notion and notation of congruence to good effect in the *Disquisitiones*. He clarified many known results, such as Fermat's little theorem and Euler's criterion, and he gave the first proofs of results Euler, Lagrange, and Legendre had attacked without success, such as quadratic reciprocity and the existence of primitive roots. He also gave the very neat proof of Wilson's theorem we used in Section 6.5. Flushed with his success, he made the following remarks about the theorem and its history:

> It was first published by Waring and attributed to Wilson: Waring *Meditationes Algebraicae* (3rd ed., Cambridge, 1782, p. 380). But neither of them was able to prove the theorem, and Waring confessed that the demonstration seemed more difficult because no *notation* can be devised to express a prime number. But in our opinion truths of this kind should be drawn from notions rather than notations. (Gauss (1801), article 76.)

He then proceeded to give his proof, with the help of congruence notation of course.

With hindsight, we can see that the congruence *notion* is implicit in many results that were known before Gauss. A simple example is the rule of casting out nines, and more sophisticated examples are Fermat's little theorem and Wilson's theorem. The latter theorem is another of the "universal" theorems, having been discovered at least twice before Wilson. Leibniz stated it in an unpublished paper around 1670, and its first known appearance is in a work of the Arab mathematician and scientist Abū 'Ali al-Hasan ibn al-Haytham (965–1039). The first known *proof*, however, is the one given by Lagrange in 1771.

The concept of congruence mod n, and particularly Gauss's notation for it, makes such results easier to discover and prove by clearing the page of all multiples of n. Having to write $\ldots \equiv \ldots$ (mod n) rather than $\ldots = \ldots$ is a small price to pay for the simplification, because congruences can be manipulated like equations anyway.

Moreover, if numbers are replaced by their congruence classes, as in Section 6.3, then congruence of numbers is replaced by equality of their congruence classes, and hence we can work with equations after all. The price to pay in this case is accepting *classes* as mathematical objects, like numbers.

Congruence classes were introduced by Dedekind in 1857, the year before he proposed the more radical idea of defining real numbers as pairs of sets of rationals. In the 1870s he used sets again to give meaning to other notions that until then had only a ghostly existence—the idea of an "ideal algebraic number" and the idea of a "point on a Riemann surface." His contemporaries found these ideas *too* radical, and it took several decades of exposure before mathematicians accepted sets as mathematical objects and realized that they made life simpler.

Rings, Fields, and Abelian Groups

Like the congruence concept, rings and fields were implicit in number theory long before they became explicit. In fact, it was only around 1900 that the ring concept was recognized at all, partly because it took that long to recognize that ordinary integers and congruence classes had a lot in common. Writing down what they have in common with each other (and with other "integer-like" objects, such as polynomials), mathematicians arrived at what we called the *ring properties* in Section 1.4. The field concept was recognized in a similar way, by writing down the common properties of various sets of objects for which $+, -, \times$, and $\div$ are meaningful—rational numbers, congruence classes mod p, and rational functions, for example.

The power of an abstract concept, like that of a field, is that it allows us to treat some outlandish mathematical objects like old friends. For example, it tells us that polynomials with mod p coefficients behave the same as ordinary polynomials with rational or real coefficients. This is why Lagrange's polynomial theorem is essentially the same as the corresponding theorem about ordinary polynomials; they both depend only on the fact that the coefficients belong to a field.

Another abstraction that number theory pushes into the limelight is the concept of an *abelian group*. This concept is actually simpler, and hence more general, than the concept of ring or field. A ring involves two operations, + and ×, but an abelian group involves only one, usually written + but sometimes · or ×. If the group operation is written as +, the abelian group properties are the ring properties of +:

$$a + (b + c) = (a + b) + c \qquad \text{(associative law)}$$
$$a + b = b + a \qquad \text{(commutative law)}$$
$$a + (-a) = 0 \qquad \text{(inverse property)}$$
$$a + 0 = a \qquad \text{(identity property)}.$$

To be precise, an abelian group is a set A with an operation +, an *identity element* called 0, and for each a in A an *inverse of a*, written $-a$, with the four properties just given. The notation with + as the group operation, 0 as the identity, and $-a$ as the inverse of a is called *additive notation*. Naturally, it is used for groups where the operation is ordinary addition, or something related to it such as addition of congruence classes.

There is also a *multiplicative notation*, in which the group operation is called · or ×, the identity is called 1, and the inverse of a is called a^{-1}.

In multiplicative notation, the abelian group properties are

$$a \times (b \times c) = (a \times b) \times c \qquad \text{(associative law)}$$
$$a \times b = b \times a \qquad \text{(commutative law)}$$
$$a \times a^{-1} = 1 \qquad \text{(inverse property)}$$
$$a \times 1 = a \qquad \text{(identity property)}.$$

Multiplicative notation is natural for groups like the nonzero rationals, where $\times$ is ordinary multiplication, or groups with a related "multiplication," like the nonzero congruence classes mod p. The *abelian* property, by the way, is the commutative law. If this property is dropped, we have what is simply called a *group*. Nonabelian groups include the groups of transformations occurring in geometry (see Section 3.8*). Thus the general group concept unifies ideas from both geometry and number theory and helps to explain the deep connections between the two.

The commutative property of abelian groups makes them easier to handle than general groups, so group theory tends to be easier in number theory than geometry. In fact, many of the groups in number theory are of a specially simple type called *cyclic* groups.

In additive notation, a cyclic group C consists of the elements

$$\ldots, -2, -1, 0, 1, 2, \ldots.$$

If C is infinite it is necessarily the integers $\mathbb{Z}$ under ordinary addition. If C is finite, with n elements, it is necessarily $\mathbb{Z}/n\mathbb{Z}$ under addition of congruence classes. In multiplicative notation, a cyclic group looks like

$$\ldots, c^{-2}, c^{-1}, 1, c, c^2, \ldots.$$

for some element c of C. For example, $\{\ldots, 2^{-2}, 2^{-1}, 1, 2, 2^2, \ldots\}$ is an infinite cyclic subgroup of the rationals, $\mathbb{Q}$. The function $f(2^n) = n$ is an isomorphism between this group and $\mathbb{Z}$, a one-to-one correspondence that sends the product $2^m \times 2^n$ to the corresponding sum $m + n$.

An example of a finite cyclic group under $\times$ is $\{1, 2, 3, 4\}$ under mod 5 multiplication. This group also consists of the powers of 2, but now taken mod 5, because

$$1 \equiv 2^0 \pmod 5$$
$$2 \equiv 2^1 \pmod 5$$
$$3 \equiv 2^3 \pmod 5$$
$$4 \equiv 2^2 \pmod 5.$$

More generally, $\{1, 2, 3, \ldots, p-1\}$ is a cyclic group under mod p multiplication for any prime p. This far-from-obvious result follows from the existence of primitive roots, mentioned in Section 6.7.

Applied Number Theory

Fermat's little theorem lay buried in the number theory books for more than 300 years before starting a new life as a fundamental tool of espionage and commerce. This transformation from pure to applied (or is it clean to dirty?) was brought about by the discovery of the RSA *public key cryptosystem* in 1977. Named after its authors, Rivest, Shamir, and Adleman, RSA is a simple method for encoding and decoding messages based on Fermat's little theorem.

Like many traditional codes, RSA scrambles and unscrambles a message using a *key*, a long sequence of digits known only to sender and receiver. Its novel feature is that the sender needs to know only part of the key, which can therefore be made public; only the receiver needs to know the whole key. The receiver's key is in fact a pair (p_1, p_2) of large prime numbers (around 100 digits each), while the public key is their product $p_1 p_2$. The theory behind the system is now explained in most number theory textbooks, for example Niven, Zuckerman, and Montgomery (1991).

The reason the product $p_1 p_2$ is effectively "less information" than the pair of factors p_1, p_2 is that there is no known method for factorizing a random product of 100 digit primes in reasonable time. Although p_1 and p_2 can in principle be derived from $p_1 p_2$, in practice they cannot, and RSA remains a secure system as long as factorization remains hard. A lot of money is riding on the assumption that it will *always* be hard. An industry has sprung up supplying easy-to-use RSA systems and accessories, in some cases even offering large primes for sale! In turn, this has stimulated much research on the problems of factorization and prime recognition.

Fermat's little theorem is fundamental to this research, because it gives an easy way to recognize when a number is *not* prime. The argument goes as follows. If p is prime and $1 < a < p$, then

$$a^{p-1} \equiv 1 \pmod{p}$$

by Fermat's little theorem. It follows that a number n is *not* prime if

$$a^{n-1} \not\equiv 1 \pmod{n}$$

for some a between 1 and n. We then call a a *witness* that n is not prime. When n is large, finding a witness is generally much easier than finding a divisor of n, because a^{n-1} mod n can be quickly computed by the repeated squaring method (Exercises 6.2.6 and 6.2.7*), and 2 or 3 is usually a witness. There are rare cases where a witness does not exist, but the method extends to cover these cases without greatly increasing the computing time.

An excellent introduction to these aspects of number theory may be found in Chapter 33 of Cormen, Leiserson, and Rivest (1990). It is particularly interesting to observe that most of the fundamentals of pure number theory are needed: Euclidean algorithm, abelian groups, Euler's phi function, Chinese remainder theorem, and Fermat's little theorem.

7

CHAPTER

Complex Numbers

7.1 Addition, Multiplication, and Absolute Value

Complex numbers are objects of the form $a + b\sqrt{-1}$, where a and b are real numbers and $\sqrt{-1}$ is... what? Mathematicians worried about this question for several centuries and did not come up with a good answer until the 19th century, by which time complex numbers had become indispensable in virtually all fields of mathematics. Their story is perhaps the supreme illustration of a saying of Hilbert's: "In mathematics, existence means freedom from contradiction."[1] Mathematicians came to believe in complex numbers because they worked, not because they could define them, and finding a definition was not a high priority until *all* concepts of number came under scrutiny.

So let us begin by assuming there is such a thing as $i = \sqrt{-1}$, and see where this leads. As we did when we introduced other new numbers, such as the integers and the reals, we want to retain the

---

[1] See Constance Reid's *Hilbert*, p. 98.

properties of the old numbers as far as possible. We therefore assume i is like any other number, except that $i^2 = -1$. Addition of complex numbers does not even involve i^2, so it is completely straightforward:

$$(a_1 + ib_1) + (a_2 + ib_2) = (a_1 + a_2) + i(b_1 + b_2).$$

If we strip each complex number $a + ib$ down to its essence, the *ordered pair* of real numbers (a, b), then addition of complex numbers is simply separate addition of a and b components:

$$(a_1, b_1) + (a_2, b_2) = (a_1 + a_2, b_1 + b_2), \qquad (\text{+ rule})$$

as one does with direct products (Section 6.6). The a and b components are traditionally called the *real* and *imaginary* parts of $a + ib$.

The interesting properties of complex numbers begin with multiplication, where $i^2 = -1$ becomes involved:

$$(a_1 + ib_1)(a_2 + ib_2) = (a_1a_2 - b_1b_2) + i(b_1a_2 + a_1b_2).$$

In terms of ordered pairs, multiplication is the rule

$$(a_1, b_1)(a_2, b_2) = (a_1a_2 - b_1b_2, b_1a_2 + a_1b_2) \qquad (\times \text{ rule}).$$

This rule is more mysterious, but it gives us a cheap way to define the complex numbers without worrying about $\sqrt{-1}$: simply define them to be ordered pairs of reals (a, b) with addition and multiplication defined by the + and × rules just given. This was first done by Hamilton in 1833.

Perhaps it seems underhand to define multiplication by the × rule when one knows it is just the disguised result of assuming $i^2 = -1$, but it isn't! The × rule was in use long before anyone dreamt of $\sqrt{-1}$. The first hint of it appears in Diophantus' *Arithmetica*, Book III, Problem 19, where he says:

> 65 is naturally divided into two squares in two ways, namely into $7^2 + 4^2$ and $8^2 + 1^2$, which is due to the fact that 65 is the product of 13 and 5, each of which is the sum of two squares.

Apparently he knew that the product of sums of squares is itself a sum of squares, in two ways, which points to the identity:

$$(a_1^2 + b_1^2)(a_2^2 + b_2^2) = (a_1a_2 \pm b_1b_2)^2 + (b_1a_2 \mp a_1b_2)^2.$$

(He had the special case $a_1 = 3$, $b_1 = 2$, $a_2 = 2$, $b_2 = 1$.) This remarkable identity was first observed explicitly by Abū Ja'far al-Khazin around 950 A.D., commenting on this problem of Diophantus, and it was proved in Fibonacci's *The Book of Squares* in 1225.

Diophantus talks about products of sums of squares, but he views $a^2 + b^2$ as the square on the hypotenuse of the right-angled triangle with sides a and b. His view reveals an important aspect of the *geometric interpretation of complex numbers*, which gradually emerged during the 16th, 17th, and 18th centuries and became standard in the 19th. A second aspect, which virtually completes the picture, will be discussed in the next section. For the moment, let us see how much is visible from Diophantus' viewpoint.

The triangle with sides a and b represents the *pair* (a, b), and hence corresponds to what we would call the complex number $a + ib$. We interpret $a + ib$ as the *vertex* of a triangle in the plane with one vertex at the origin and sides a and b parallel to the axes (Figure 7.1), and interpret the set $\mathbb{C}$ of all complex numbers as the plane. The "hypotenuse" $\sqrt{a^2 + b^2}$ of the triangle with sides a and b is what we call the *absolute value* $|a + ib|$ of the corresponding complex number $a + ib$. Diophantus' identity says that this geometrically defined quantity has a simple algebraic property; it is *multiplicative*. The absolute value of a product is the product of the absolute values:

$$|a_1 + ib_1||a_2 + ib_2| = |(a_1 + ib_1)(a_2 + ib_2)|.$$

In fact, the two sides of this equation are just the square roots of the two sides of Diophantus' identity, with the lower signs chosen on the right-hand side.

Of course, there was no reason for Diophantus to speak of a "product of triangles," or even to think of it. All he wanted was a *rule* for taking two triangles and producing a third for which the hypotenuse was the product of the two hypotenuses he started with. Applied

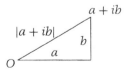

FIGURE 7.1   Geometric meaning of absolute value.

to triangles representing pairs (a_1, b_1) and (a_2, b_2), however, his rule produced what we call their product, $(a_1a_2 - b_1b_2, b_1a_2 + a_1b_2)$. (The rule for producing the other triangle in Diophantus' identity can also be interpreted as a product of complex numbers; see the exercises.) It is all the more surprising that Diophantus devised this rule to solve problems about integers. As we shall see in the next section, its geometric significance goes far deeper than the interpretation of absolute value as a hypotenuse.

Remark. The product of triangles is closely related to the parameterization of Pythagorean triples (which Diophantus used in the same problem, by the way, so he was probably aware of a connection). *The triangle with sides $(u^2 - v^2, 2uv)$ and hypotenuse $u^2 + v^2$ is the "square" of the triangle with sides (u, v) with hypotenuse $\sqrt{u^2 + v^2}$.* This is an easy calculation with complex numbers:

$$(u + iv)^2 = u^2 - v^2 + 2iuv.$$

Exercises

Proving Diophantus' identity is not as hard as discovering it in the first place; just expand both sides and compare them. But if complex numbers are already familiar, the identity may be discovered by factorizing the product $(a_1^2 + b_1^2)(a_2^2 + b_2^2)$ and recombining the factors in a different way. This was done by Euler (1770).

7.1.1. Give a derivation of Diophantus' identity by suitably combining the factors in $(a_1 + ib_1)(a_1 - ib_1)(a_2 + ib_2)(a_2 - ib_2)$.

An interesting generalization of Diophantus' identity was discovered by the Indian mathematician Brahmagupta around 600 A.D.:

$$(a_1^2 - db_1^2)(a_2^2 - db_2^2) = (a_1a_2 + db_1b_2)^2 - d(a_1b_2 + a_2b_1)^2.$$

7.1.2. Give a derivation of Brahmagupta's identity by suitably grouping the factors in $(a_1 + \sqrt{d}b_1)(a_1 - \sqrt{d}b_1)(a_2 + \sqrt{d}b_2)(a_2 - \sqrt{d}b_2)$.

The identities of Diophantus and Brahmagupta are, of course, valid for all real values of $a_1, b_1, a_2, b_2,$ and d. However, they are of most interest when these values are integers. In that case, they show that *the product of two integers of the form $a^2 - db^2$ is another integer of the same form.* This discovery is the beginning of a very long story we shall take up in Section 7.6.

Now let us return to the second Diophantus identity, with the signs switched.

7.1.3.   To which complex numbers should $a_1^2 + b_1^2$ and $a_2^2 + b_2^2$ be attached, for

$$(a_1^2 + b_1^2)(a_2^2 + b_2^2) = (a_1 a_2 + b_1 b_2)^2 + (b_1 a_2 - a_1 b_2)^2$$

to express the multiplicative property of absolute value?

7.1.4.   Find a second form of Brahmagupta's identity, also with signs switched.

7.2   Argument and the Square Root of −1

The product of triangles that was implicit in Diophantus became explicit in Viète's *Genesis triangulorum* around 1590. He actually drew diagrams of the right-angled triangles (a_1, b_1) and (a_2, b_2) and their two products, similar to Figure 7.2 but without labeled angles or the × and = signs.

Viète was interested in the shape of the triangles more than the length of their hypotenuses, and this led him to a wonderful discovery: the product of triangles produces not only the product of hypotenuses, but the *sum* of angles. The angles that are added are those shown in Figure 7.2. In fact, the ratio of the sides in the

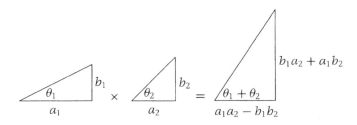

FIGURE 7.2   The first product of triangles.

product triangle is precisely what we get from the addition formula for tan (Section 5.4):

$$\tan(\theta + \phi) = \frac{\tan\theta + \tan\phi}{1 - \tan\theta\,\tan\phi}$$

$$= \frac{\frac{b_1}{a_1} + \frac{b_2}{a_2}}{1 - \frac{b_1}{a_1}\frac{b_2}{a_2}}$$

$$= \frac{b_1 a_2 + a_1 b_2}{a_1 a_2 - b_1 b_2}.$$

Like Diophantus, Viète was thinking about triangles, not complex numbers. Nevertheless, just as Diophantus observed the multiplicative property of what we call the absolute value, Viète observed the *additive* property of what we call the *argument* of the complex number $a + ib$, the angle θ with $\cos\theta = a/\sqrt{a^2 + b^2}$ and $\sin\theta = b/\sqrt{a^2 + b^2}$. (The only limitation to interpreting complex numbers $a + ib$ as triangles is that a and b cannot be negative or zero. This does mean, however, that there is no interpretation of the crucial object i.)

When a complex number $a + ib$ is viewed as a point (a, b) of the plane, its absolute value and argument are its *polar coordinates* $r = \sqrt{a^2 + b^2}$ and θ (Figure 7.3). The multiplicative property of absolute value and the additive property of argument give the product of complex numbers in polar coordinates:

$$(a_1 + ib_1)(a_1 + ib_2) = r_1(\cos\theta_1 + i\sin\theta_1)r_2(\cos\theta_2 + i\sin\theta_2)$$

$$= r_1 r_2\,(\cos(\theta_1 + \theta_2) + i\sin(\theta_1 + \theta_2)).$$

Because (a, b) is completely determined by r and θ, this is equivalent to the $\times$ rule in the previous section as a definition of product. It shows multiplication in a much more geometric light, and it gives

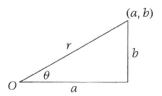

FIGURE 7.3   Absolute value and argument as polar coodinates.

a geometric interpretation of multiplication by $i = \sqrt{-1}$. Because $i = \cos\frac{\pi}{2} + i\sin\frac{\pi}{2}$, *multiplication by i adds $\pi/2$ to the argument of each complex number.* That is, *it rotates the plane of complex numbers counterclockwise through $\pi/2$.*

With hindsight, it is natural for multiplication by i to be a quarter turn. After all, multiplication by i twice is multiplication by -1, which is a half turn of the real number line. Algebraically speaking, multiplication is an operation of *period* 4, because the powers of i recur every four steps:

$$1, \quad i, \quad -1, \quad -i, \quad 1, \quad i, \quad -1, \quad -i, \quad 1, \quad \ldots .$$

Hence $i = \sqrt{-1}$ can only exist in a system containing operations of period 4. Such a system need not be the full set of complex numbers. For example, it could be the *Gaussian integers*, the set of numbers $a + ib$ where a and b are integers. We shall study these in section 7.4. The system can even be the finite field $\mathbb{Z}/p\mathbb{Z}$ for a suitable value of p. As we saw in Exercises 6.5.8 and 6.5.9, -1 is a square in $\mathbb{Z}/p\mathbb{Z}$ just in case $p = 4n + 1$, and it follows from the existence of primitive roots (mentioned in the exercises for Section 6.7) that this happens precisely when $\mathbb{Z}/p\mathbb{Z}$ has elements of period 4.

More of a surprise is that i, the fourth root of 1, together with the real numbers, gives a nontrivial nth root of 1 for *all* natural numbers n. In fact, it follows from the additive property of argument that

$$\left(\cos\frac{2\pi}{n} + i\sin\frac{2\pi}{n}\right)^n = \cos 2\pi + i\sin 2\pi = 1,$$

so $\cos\frac{2\pi}{n} + i\sin\frac{2\pi}{n}$ is a nontrivial nth root of 1. (Nontrivial, because it is not equal to 1 itself.) The *fundamental theorem of algebra*, whose proof is beyond the scope of this book, says that much more is true: *any equation $p(x) = 0$, where p is a polynomial with real coefficients, has a solution in the set $\mathbb{C}$ of complex numbers.*

Exercises

The complex numbers are such a fundamental part of mathematics that it is no wonder that aspects of them (such as the multiplicative property of absolute value) were glimpsed long before they were recognized as numbers. The latter possibility could not arise until there was reason to add *and* multiply them. This did not happen until the 16th century, when a number of Italian mathematicians discovered how to solve cubic equations. It turned out, for example, that the solutions of

$$x^3 = px + q$$

are given by the *Cardano formula*:

$$x = \sqrt[3]{\frac{q}{2} + \sqrt{\left(\frac{q}{2}\right)^2 - \left(\frac{p}{3}\right)^3}} + \sqrt[3]{\frac{q}{2} - \sqrt{\left(\frac{q}{2}\right)^2 - \left(\frac{p}{3}\right)^3}}.$$

7.2.1. Substitute $u + v$ for x in $x^3 = px + q$, and deduce that $u + v$ will satisfy this equation if $3uv = p$ and $u^3 + v^3 = q$.

7.2.2. Substitute $v = p/3u$ in $u^3 + v^3 = q$ and solve the resulting quadratic in u^3. Deduce that

$$u^3, v^3 = \frac{q}{2} \pm \sqrt{\left(\frac{q}{2}\right)^2 - \left(\frac{p}{3}\right)^3}$$

and explain the Cardano solution of $x^3 = px + q$.

This is all very well, but does it account for the obvious solution $x = 4$ of $x^3 = 15x + 4$?

7.2.3. Show that, according to the Cardano formula, the solutions of $x^3 = 15x + 4$ are

$$x = \sqrt[3]{2 + 11\sqrt{-1}} + \sqrt[3]{2 - 11\sqrt{-1}}.$$

Rafael Bombelli (1572) had a hunch that this apparent conflict might be resolved as follows. He guessed there was an n with

$$\sqrt[3]{2 + 11\sqrt{-1}} = 2 + n\sqrt{-1},$$

$$\sqrt[3]{2 - 11\sqrt{-1}} = 2 - n\sqrt{-1},$$

so the solution $x = 4$ could result from cancelation of imaginary terms $n\sqrt{-1}$ and $-n\sqrt{-1}$. This hunch turned out to be correct.

7.2.4. Show that $(2 + \sqrt{-1})^3 = 2 + 11\sqrt{-1}$ and $(2 - \sqrt{-1})^3 = 2 - 11\sqrt{-1}$.

These calculations, which involve both addition and multiplication of complex numbers, were enough to convince many mathematicians that complex numbers were subject to the same laws as the reals. Still, whenever possible one made an independent check using real numbers alone.

Viète found an interesting way to do this for cubic equations in cases where the Cardano formula leads to square roots of negative numbers. He used the formula

$$\cos 3\theta = 4\cos^3 \theta - 3\cos\theta$$

and coaxed the equation $x^3 - px = q$ into the form $4y^3 - 3y = c$, where c could be set equal to $\cos 3\theta$ for some angle θ. Then $y = \cos\theta$ is a solution of $4y^3 - 3y = c$, and x is easily found from y.

7.2.5. Find a substitution $x = ky$ so that $2x^3 = 3x - 1$ becomes $4y^3 - 3y = -1/\sqrt{2} = \cos\frac{3\pi}{4}$, and hence find a solution of $2x^3 = 3x-1$ by Viète's method.

(You will probably only find the obvious solution, but see whether it is so obvious from Cardano's formula.)

7.3   Isometries of the Plane

Rotation of the plane about O through $\pi/2$ is one of the isometries of the plane we studied in Section 3.6. The discovery that this rotation is simply multiplication by the complex number i prompts us to look at isometries again. It seems as though they can be very concisely described in terms of complex numbers.

We interpret the Euclidean plane as the set $\mathbb{C}$ of complex numbers, so isometries are certain functions on $\mathbb{C}$. Translations are the easiest to grasp. Translating each point $x + iy$ by a in the x-direction and b in the y-direction is the same as *adding* $a + ib$ to $x + iy$, so the translation function $\text{tran}_{a,b}$ is the function of a complex variable $z = x + iy$ defined by

$$\text{tran}_{a,b}(z) = z + a + ib.$$

By the additive property of the argument, rotation about O through θ is multiplication by $\cos\theta + i\sin\theta$, hence

$$\text{rot}_{O,\theta}(z) = (\cos\theta + i\sin\theta)z.$$

Because $\cos^2 \theta + \sin^2 \theta = 1$, this is multiplication of z by a complex number of absolute value 1. Conversely, any complex number c of absolute value 1 is of the form $\cos \theta + i \sin \theta$; in fact, θ is the argument of c. Thus rotation about O is multiplication by a fixed complex number of absolute value 1.

Rotation about any point $P = (u, v)$, through angle θ, is the composite of three functions:

- $\text{tran}_{-u, -v} =$ translation of P to O,

- $\text{rot}_{O, \theta} =$ rotation about O through θ,

- $\text{tran}_{u, v} =$ translation of O back to P.

Thus any rotation may be composed from the functions $\text{tran}_{a, b}$ and $\text{rot}_{O, \theta}$.

In the exercises to Section 3.8* it was shown that translations and rotations together are all the products of an even number of reflections, and that they also are the orientation-preserving isometries of the plane. Complex functions give another very neat way to describe them, without assuming these previous results.

Characterization of translations and rotations. *The translations and rotations of the Euclidean plane are the complex functions of the form*

$$f(z) = cz + d,$$

where c is a complex number with $|c| = 1$ and d is an arbitrary complex number.

Proof We know that translations and rotations about O are of the required form. Also, if $f_1(z) = c_1 z + d_1$ and $f_2(z) = c_2 z + d_2$ are of the required form, then so is

$$f_1 f_2(z) = c_1(c_2 z + d_2) + d_1 = (c_1 c_2)z + (c_1 d_2 + d_1),$$

because $|c_1 c_2| = |c_1||c_2| = 1 \times 1 = 1$ by the multiplicative property of absolute value. Thus any composite of translations and rotations about O, which we know includes rotations about arbitrary points, is of the form $f(z) = cz + d$ with $|c| = 1$.

Conversely, suppose we are given a function $f(z) = cz + d$ with $|c| = 1$. If $c = 1$, then we simply have the translation $f(z) = z + d$. If not, consider the rotation about the point e obtained by composing

$$f_1(z) = z + e,$$
$$f_2(z) = cz,$$
$$f_3(z) = z - e,$$

where e remains to be determined. We have

$$f_1 f_2 f_3(z) = c(z - e) + e = cz - ce + e,$$

which equals the given function if

$$d = -ce + e = e(1 - c);$$

that is, if

$$e = \frac{d}{1 - c}.$$

By hypothesis, $1 - c \neq 0$, so we can find the point e, and $f(z) = cz + d$ is a rotation about it. $\square$

Exercises

Another important isometry is reflection in the x-axis, the function that sends $x + iy$ to its *conjugate* $x - iy$. The conjugate of z is denoted by $\bar{z}$ and has the following easily checked properties.

7.3.1. Check that $\overline{z_1 + z_2} = \overline{z_1} + \overline{z_2}$, $\overline{z_1 z_2} = \overline{z_1}\,\overline{z_2}$ and $z\bar{z} = |z|^2$.

Composing conjugation with translations and rotations gives a further class of isometries

$$\bar{f}(z) = c\bar{z} + d \quad \text{where } |c| = 1.$$

These, together with the functions $f(z) = cz + d$ already found, make up *all* isometries of the Euclidean plane. A quick way to prove this is to combine the characterization of translations and rotations with the three reflections theorem and related results in Section 3.6. By Exercises 3.6.3 and 3.6.4, the composite of two reflections is a translation or rotation, hence of the form $f(z) = cz + d$ with $|c| = 1$.

7.3.2. Suppose a Euclidean isometry g is a composite of one or three reflections. Show that $\overline{g(z)}$ is a rotation or translation.

7.3.3. Deduce from Exercise 7.3.2 that $g(z) = \overline{\overline{g(z)}}$ is of the form $c\bar{z} + d$ with $|c| = 1$.

7.3.4. Conclude from the preceding results that the Euclidean isometries are precisely the functions $f(z) = cz + d$ and $\bar{f}(z) = c\bar{z} + d$ with $|c| = 1$.

This characterization makes it easy to see the difference between

1. the orientation-preserving isometries, which are those of the form $f(z) = cz + d$ (because these are the translations and rotations), and

2. the orientation-reversing isometries, which are those of the form $\bar{f}(z) = c\bar{z} + d$ (because these are the rest).

It also gives an easier way to prove the result of Exercise 3.6.6*, that any orientation-reversing isometry is a glide reflection. The idea is to rotate and translate the coordinate system until the isometry looks like $\bar{h}(z) = \bar{z} + a$, with a real, which is a glide reflection along the x-axis.

7.3.5. Show that if $z' = (\cos\phi + i\sin\phi)z$ is taken as the new coordinate of the point z, then the x'- and y'-axes of this new coordinate system are the result of rotating the x- and y-axes through $-\phi$.

Now suppose that $\bar{f}(z) = (\cos\theta + i\sin\theta)\bar{z} + d$ is an orientation-reversing isometry. Thus, in the old coordinate system, the isometry sends z to $(\cos\theta + i\sin\theta)\bar{z} + d$.

7.3.6. If $z' = (\cos(-\theta/2) + i\sin(-\theta/2))z$, show that the point with new coordinate z'

- has old coordinate $(\cos(\theta/2) + i\sin(\theta/2))z'$,

- which is sent to the point with old coordinate

$$(\cos(\theta/2) + i\sin(\theta/2))\bar{z'} + d,$$

- which has new coordinate $\bar{z'} + d$.

Conclude that the isometry is given by $\bar{g}(z') = \bar{z'} + d$ in the new coordinate system.

Finally, suppose that $d = a + ib$, where a and b are real. Replace the coordinate z' (which is now called the *old coordinate*) by a new coordinate z'' defined by $z'' = z' - ib/2$.

7.3.7. Show the point with new coordinate z''

- has old coordinate $z'' + ib/2$,

- which is sent to the point with old coordinate $\overline{z''} + a + ib/2$,

- which has new coordinate $\overline{z''} + a$.

Conclude that the isometry is given by $\overline{h}(z'') = \overline{z''} + a$ in the new coordinate system and hence is a glide reflection.

In case you are wondering about functions of the form $f(z) = cz + d$, where c is *not* required to have absolute value 1, see the following.

7.3.8. Show that any function of the form $f(z) = cz + d$, where $c \neq 0$, is a composite of a translation or rotation with a *dilatation* —a function of the form $g(z) = rz$ where r is real.

As was mentioned in Section 3.10, these functions are precisely the *similarities*, and they are the only mappings of the plane that preserve angles.

7.4 The Gaussian Integers

In the complex numbers, the counterparts of the integers are called the *Gaussian integers*. They are the complex numbers of the form $a + ib$ where a and b are in $\mathbb{Z}$, and the set of them is denoted by $\mathbb{Z}[i]$. Like $\mathbb{Z}$, $\mathbb{Z}[i]$ is a ring and has notions of divisor and prime. For this reason alone, it is interesting to investigate the arithmetic of $\mathbb{Z}[i]$, but even more interesting is the insight it gives into $\mathbb{Z}$ itself. In a sense, $\mathbb{Z}[i]$ refines our understanding of $\mathbb{Z}$ by allowing ordinary integers to be analyzed in finer detail.

A simple example is the Diophantus identity

$$(a_1^2 + b_1^2)(a_2^2 + b_2^2) = (a_1 a_2 - b_1 b_2)^2 + (b_1 a_2 + a_1 b_2)^2$$

for a_1, b_1, a_2, and b_2 in $\mathbb{Z}$. As already suggested in Exercise 7.1.1, this identity is more understandable in $\mathbb{Z}[i]$, where we have the factorizations

$$a_1^2 + b_1^2 = (a_1 + ib_1)(a_1 - ib_1),$$
$$a_2^2 + b_2^2 = (a_2 + ib_2)(a_2 - ib_2).$$

If we rearrange these factors of $(a_1^2 + b_1^2)(a_2^2 + b_2^2)$ as

$$(a_1 + ib_1)(a_2 + ib_2)(a_1 - ib_1)(a_2 - ib_2)$$

and then combine the first two and the last two, we get

$$[(a_1a_2 - b_1b_2) + i(b_1a_2 + a_1b_2)][(a_1a_2 - b_1b_2) - i(b_1a_2 + a_1b_2)],$$

which is a Gaussian integer factorization of $(a_1a_2 - b_1b_2)^2 + (b_1a_2 + a_1b_2)^2$.

We noted in Section 7.1 that Diophantus' identity shows that the absolute value function $|a + ib|$ is multiplicative. Even more directly, it shows that the function $|a + ib|^2$ is multiplicative. The latter is a very useful function on $\mathbb{Z}[i]$, called the *norm* of $a + ib$ and written $N(a + ib)$. Diophantus' identity is precisely the *multiplicative property of the norm*:

$$N((a_1 + ib_1)(a_2 + ib_2)) = N(a_1 + ib_1)N(a_2 + ib_2).$$

The norm is useful because:

- It is an ordinary integer and hence reduces some questions about $\mathbb{Z}[i]$ to questions about $\mathbb{Z}$.

- It is multiplicative and hence the norm of a factor divides the norm of a product.

In particular, the norm draws our attention to the *units* $1, -1, i, -i$ of $\mathbb{Z}[i]$, the members of norm 1. These are the numbers that divide every Gaussian integer and hence can be regarded as redundant factors (like 1 and -1 in $\mathbb{Z}$). When unit factors are disregarded, each Gaussian integer can be split into finitely many factors

$$a + ib = (a_1 + ib_1)(a_2 + ib_2) \cdots (a_k + ib_k),$$

which are *Gaussian primes* in the sense that $a_j + ib_j$ has no divisors of smaller norm except units. It follows that $a_j + ib_j$ has no divisors at all except units and multiples of itself by units.

Gaussian prime factorizations exist in $\mathbb{Z}[i]$ for much the same reason that prime factorizations exist in $\mathbb{Z}$: *each Gaussian integer has a Gaussian prime divisor* (compare with Section 1.3). If $a + ib$ has no nonunit divisor of smaller norm, then $a+ib$ itself is a Gaussian prime. Otherwise, take a nonunit divisor $a' + ib'$ of smaller norm, and see whether $a' + ib'$ has a nonunit divisor $a'' + ib''$ of still smaller norm, and so on. Because the norms are natural numbers, this process ends in a finite number of steps, necessarily with a Gaussian prime divisor $a_1 + ib_1$. We then repeat the process on the Gaussian integer $(a + ib)/(a_1 + ib_1)$, which has smaller norm than $a + ib$, and so on.

Exercises

The norm sometimes enables us to recognize Gaussian primes.

7.4.1. Find some Gaussian integers whose norms are prime.

7.4.2. A Gaussian integer with prime norm is a Gaussian prime. Why?

However, ordinary primes are not necessarily Gaussian primes.

7.4.3. Show that 2 is not a Gaussian prime. Also find an odd prime that is not a Gaussian prime.

Your odd prime should be of the form $4n+1$, because ordinary primes of the form $4n + 3$ *are* Gaussian primes. This is proved with the help of conjugation.

7.4.4. Suppose that p is an ordinary prime and $p = (a + ib)c$ is a Gaussian factorization without units. Show in turn that

- $p = (a - ib)\bar{c}$
- $p^2 = (a^2 + b^2)|c|^2$
- $p = a^2 + b^2$
- p is not of the form $4n + 3$.

In Exercise 1.3.5 it was proved that there are infinitely many primes of the form $4n + 3$, so it follows from Exercise 7.4.4 that there are infinitely many Gaussian primes. The same result can also be proved directly, in the manner of Euclid, once we clarify the idea of division with remainder in $\mathbb{Z}[i]$. This will be done in the next section.

However, before going more deeply into $\mathbb{Z}[i]$, it should be pointed out that $\mathbb{Z}[i]$ is not the only ring of "integers" in $\mathbb{C}$. Unlike $\mathbb{R}$, which has $\mathbb{Z}$ as its only integers, $\mathbb{C}$ has many subrings that can reasonably be regarded as integers. Another example is the set

$$\mathbb{Z}[\sqrt{-2}] = \{a + b\sqrt{-2} : a, b \in \mathbb{Z}\}.$$

This set is a ring because the sum and product of any two of its members are also members, whence it inherits the ring properties from $\mathbb{C}$. As on $\mathbb{Z}[i]$, the square of the absolute value gives a norm on $\mathbb{Z}[\sqrt{-2}]$, which is integer-valued and multiplicative. The "integers" in $\mathbb{Z}[i]$ and $\mathbb{Z}[\sqrt{-2}]$ are called *quadratic integers* because they satisfy quadratic equations with rational coefficients. We shall say a little more about quadratic integers in general in Section 7.8.

7.4.5. Show that $N(a + b\sqrt{-2}) = a^2 + 2b^2$. Use this norm to show that 5 is a "prime" in $\mathbb{Z}[\sqrt{-2}]$, and that 1 and -1 are the only units in $\mathbb{Z}[\sqrt{-2}]$.

7.5   Unique Gaussian Prime Factorization

We have now come to the point where further progress in the arithmetic of $\mathbb{Z}[i]$ depends on a uniqueness theorem for Gaussian prime factorization. At the same point in ordinary arithmetic (Section 1.6), we derived unique prime factorization from the fact that $\gcd(a, b) = ma + nb$ for some integers m and n, which follows in turn from the fact that $\gcd(a, b)$ is obtainable by the Euclidean algorithm.

The same argument applies in $\mathbb{Z}[i]$, except that there is no subtraction form of the Euclidean algorithm. We have to use division with remainder, which depends on the following.

Division property of $\mathbb{Z}[i]$.   *If α and β are Gaussian integers with $\beta \neq 0$, then there are Gaussian integers μ and ρ with*

$$\alpha = \mu\beta + \rho \quad and \quad N(\rho) < N(\beta).$$

Proof   Because the norm N is the square of the absolute value, it suffices to find ρ with $\alpha = \mu\beta + \rho$ and $|\rho| < |\beta|$.

Consider the set of all Gaussian integer multiples of β. The points in this set lie at the corners of a grid of squares, namely the translates by multiples of β of the square with corners 0, β, $i\beta$, and $(1 + i)\beta$. (Figure 7.4; the grid is square because multiplication by i rotates through a right angle.)

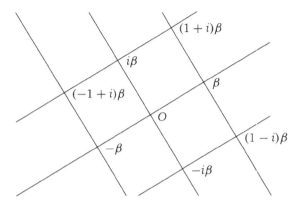

FIGURE 7.4 Multiples of a Gaussian integer.

Let $\mu\beta$ be the corner nearest to α, so $|\alpha - \mu\beta|$ is the distance between them. This distance is the hypotenuse of a right-angled triangle with sides $\leq |\beta|/2$ (Figure 7.5), hence $|\alpha - \mu\beta| < |\beta|$ by the triangle inequality. Thus if we let $\rho = \alpha - \mu\beta$ we have $\alpha = \mu\beta + \rho$ with $|\rho| < |\beta|$, as required. $\qquad\square$

Thanks to the division property, the successive divisions in the Euclidean algorithm produce remainders with strictly decreasing norms. Because the norms are natural numbers, the algorithm terminates, and it produces the gcd for the same reason as in $\mathbb{Z}$: if the algorithm starts on α and β, *all* the divisors of α and β persist as divisors of all the numbers produced by the algorithm. The gcd of α and β is not only "greatest" in the sense that all common divisors divide it; it is also greatest in norm, by the norm multiplicative property.

One can then check that the remaining steps to unique prime factorization in $\mathbb{Z}$ can be imitated (with small changes) in $\mathbb{Z}[i]$:

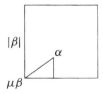

FIGURE 7.5 Distance to the nearest corner.

- $\gcd(\alpha, \beta) = \mu\alpha + \nu\beta$ for some Gaussian integers μ and ν.
- If a Gaussian prime ζ divides $\alpha\beta$, then ζ divides α or ζ divides β (Gaussian prime divisor property).
- The factors in two Gaussian prime factorizations of a Gaussian integer agree up to order and unit factors.

The latter is the "unique prime factorization theorem" for Gaussian integers.

Before drawing conclusions from this theorem, a word of caution is in order: *watch out for units!* Remember that unique prime factorization was originally proved for natural numbers, where the factorization is unique up to order. In $\mathbb{Z}[i]$, the factors can also vary up to units, and this affects some of the conclusions we can draw. In fact, this already happens in $\mathbb{Z}$, where factors can vary in sign, due to the presence of the unit -1.

Take, for example, the theorem that relatively prime numbers a and b whose product is a square are themselves squares. This is true in the natural numbers (proved in Section 4.2), but in $\mathbb{Z}$ we can conclude only that a and b are either squares or the negatives of squares. The example $(-3^2)(-5^2) = 15^2$ shows we cannot do better. Similarly, if α and β are relatively prime Gaussian integers whose product is a square, we can conclude only that each of α and β is a unit times a square. The units i and $-i$ are not squares in $\mathbb{Z}[i]$, so α and β need not be squares.

However, things get better with cubes. In $\mathbb{Z}[i]$ all the units are cubes: $1 = 1^3$, $-1 = (-1)^3$, $i = (-i)^3$, and $-i = i^3$. Thus if α and β are relatively prime Gaussian integers whose product is a cube, then α and β are not merely units times cubes, but actual cubes, because a unit times a cube is a cube.

Exercises

As mentioned in the previous set of exercises, we can use division with remainder to give a direct proof that there are infinitely many Gaussian primes.

7.5.1. If $\alpha = \mu\beta + \rho$, with $0 < |\rho| < |\beta|$, show that α is not a multiple of β.

7.5.2. Use Exercise 7.5.1 to prove that there are infinitely many Gaussian primes.

The geometric argument used to prove the division property of $\mathbb{Z}[i]$ also applies to the ring $\mathbb{Z}[\sqrt{-2}]$ discussed in the previous set of exercises.

7.5.3. Show that the multiples of a number β in $\mathbb{Z}[\sqrt{-2}]$ lie at the corners of a grid of rectangles whose sides have lengths $|\beta|$ and $\sqrt{2}|\beta|$.

7.5.4. Deduce from Exercise 7.5.3 that $\mathbb{Z}[\sqrt{-2}]$ has division property like that of $\mathbb{Z}[i]$ and hence unique prime factorization.

In the exercises to Section 1.3 we mentioned that it is not known whether there are infinitely many primes of the form $p = n^2 + 1$.

7.5.5. Show that, if $p = n^2 + 1$ is prime, then $p = (n + i)(n - i)$ is a factorization into Gaussian primes.

7.5.6. Conversely, show that if n is a natural number and $n \pm i$ are Gaussian primes, then $n^2 + 1$ is prime.

(*Hint*: Suppose $n^2 + 1$ is not prime and use unique Gaussian prime factorization.)

It follows from the last two exercises that primes of the form $n^2 + 1$ correspond to Gaussian prime pairs of the form $n \pm i$. This calls to mind the famous *twin primes problem*, which is also unsolved: are there infinitely many pairs of ordinary primes of the form $(p, p + 2)$?

7.6   Fermat's Two Squares Theorem

Now is an appropriate time to recall the words of Diophantus quoted at the beginning of this chapter:

> 65 is naturally divided into two squares in two ways, namely into $7^2 + 4^2$ and $8^2 + 1^2$, which is due to the fact that 65 is the product of 13 and 5, each of which is the sum of two squares.

When Fermat read these words, he realized that the key to representing numbers as sums of two squares was to represent *primes*. In the example, 65 is the product of the primes 5 and 13, and the two representations of 65 as sums of two squares come from the unique representations of 5 and 13 as sums of two squares by Diophantus' identity

$$(a_1^2 + b_1^2)(a_2^2 + b_2^2) = (a_1 a_2 \pm b_1 b_2)^2 + (b_1 a_2 \mp a_1 b_2)^2.$$

Fermat claimed to have proved that each prime of the form $4n + 1$ is a sum of two squares in exactly one way. However, his proof was lost, and the first known proof was given by Euler in 1749. Euler's proof was heavy going, and the theorem became a challenge to later mathematicians, to test the strength of new methods in number theory. Progressively easier and more elegant proofs were given by Lagrange, Gauss, and Dedekind. The following proof uses ideas from all three, but the crux of it is uniqueness of Gaussian prime factorization.

Fermat's two squares theorem.   *If p is a prime of the form $4n + 1$, then $p = a^2 + b^2$ for a unique pair of natural numbers a and b.*

Proof   The first step is to find a square, m^2, such that p divides $m^2 + 1$. Lagrange found a way to do this using Wilson's theorem.[2] Recall from Section 6.5 that this theorem says $(p-1)! \equiv -1 \pmod{p}$ when p is prime, so when $p = 4n + 1$ we have

$$\begin{aligned}
-1 &\equiv 1 \times 2 \times \cdots \times 4n \pmod{p} \\
&\equiv (1 \times 2 \times \cdots \times 2n)(2n+1) \times \cdots \times (4n-1) \times 4n \pmod{p} \\
&\equiv (1 \times 2 \times \cdots \times 2n)(-2n) \times \cdots \times (-2) \times (-1) \pmod{p} \\
&\quad \text{because each } p - k \equiv -k \pmod{p} \\
&\equiv (1 \times 2 \times \cdots \times 2n)^2 (-1)^{2n} \pmod{p} \\
&\equiv (1 \times 2 \times \cdots \times 2n)^2 \pmod{p}.
\end{aligned}$$

Thus if we take $m = (2n)!$ we have $-1 \equiv m^2 \pmod{p}$, and therefore p divides $m^2 + 1$.

Now $m^2 + 1$ has the Gaussian integer factorization $(m + i)(m - i)$, and p does *not* divide $m + i$ or $m - i$, because the quotients $\frac{m}{p} + \frac{i}{p}$ and $\frac{m}{p} - \frac{i}{p}$ are not Gaussian integers. It then follows from unique Gaussian prime factorization (or, more particularly, from the Gaussian prime

---

[2]The idea was indicated in Exercise 6.5.9, but to avoid dependence on the exercises, the details are given here.

divisor property) that *p is not a Gaussian prime.* Thus *p* has a Gaussian prime divisor, say, $a + ib$. Now

$a + ib$ divides $p \Rightarrow p = (a + ib)c$

for some nonunit Gaussian integer c

$\Rightarrow p = (a - ib)\bar{c}$ taking conjugates of both sides

$\Rightarrow p^2 = (a^2 + b^2)|c|^2$

multiplying preceding equations

$\Rightarrow p = a^2 + b^2$

by unique prime factorization

of natural numbers.

Conversely, if a prime $p = a^2 + b^2$, then p has the Gaussian factorization $p = (a + ib)(a - ib)$, and each factor is a Gaussian prime, because its norm is the prime p. Thus a and b are the real and imaginary parts (up to sign) of the unique Gaussian prime factors of p, and therefore $a^2 + b^2$ is the unique sum of squares equal to p. $\square$

This theorem tells us all the primes that are sums of two squares. Apart from those of the form $4n+1$, the only such prime is $2 = 1^2 + 1^2$, because primes of the form $4n + 3$ are not sums of two squares, by a congruence mod 4 argument.

The proof also tells us that if an ordinary prime p is not a Gaussian prime, then $p = a^2 + b^2$. Hence *ordinary primes of the form* $4n + 3$ *are Gaussian primes.* This information leads to the following theorem, which is closely allied with Fermat's two squares theorem and unique Gaussian prime factorization. It shows that the Gaussian primes can be regarded as "known" once the ordinary primes are known.

Classification of Gaussian primes. *Up to unit factors, the Gaussian primes are*

- *Ordinary primes of the form* $4n + 3$.

- *The factors* $a + ib$, $a - ib$ *of primes* $a^2 + b^2$ *of the form* $4n + 1$ *or* 2.

Proof By the preceding remarks, the only ordinary primes that are Gaussian primes are those of the form $4n + 3$. The factors $a + ib$ and $a - ib$ of primes $p = a^2 + b^2$ are Gaussian primes because they have norm p.

Conversely, if $a + ib$ is a Gaussian prime with $a, b \neq 0$ then so is its conjugate $a - ib$, because a nontrivial factorization of $a - ib$ would give one of $a + ib$ by conjugation. Thus Gaussian primes that are not ordinary primes come in pairs $a \pm ib$. The product $a^2 + b^2$ of such a pair is an ordinary prime, by unique Gaussian prime factorization, because a factorization of $a^2 + b^2$ into ordinary primes would be different from its Gaussian prime factorization $(a + ib)(a - ib)$. Such primes are 2 and those of the form $4n + 1$ by Fermat's two squares theorem. □

Exercises

It is possible to study the primes of $\mathbb{Z}[\sqrt{-2}]$ in a similar way, using its unique prime factorization theorem from the previous exercise set. However, to do something a little different, we shall use $\mathbb{Z}[\sqrt{-2}]$ to investigate the equation $y^3 = x^2 + 2$. Diophantus mentioned the natural number solution $x = 5$, $y = 3$ to this equation, and Fermat claimed it was the only one. The first known proof was given by Euler (1770), assuming unique prime factorization in $\mathbb{Z}[\sqrt{-2}]$ (but failing to mention it). Such a proof can be carried out rigorously as follows.

Assuming x are natural numbers with $y^3 = x^2 + 2$, note that

$$y^3 = (x + \sqrt{-2})(x - \sqrt{-2}).$$

This transforms the problem into one about cubes in $\mathbb{Z}[\sqrt{-2}]$.

7.6.1. Use congruences mod 4 to show that x is odd.

7.6.2. Deduce from Exercise 7.6.1 that $\gcd(x + \sqrt{-2}, x - \sqrt{-2}) = 1$.

Thus we have relatively prime numbers $x + \sqrt{-2}$ and $x - \sqrt{-2}$ whose product is a cube, y^3. We can conclude that $x + \sqrt{-2}$ and $x - \sqrt{-2}$ are themselves cubes in $\mathbb{Z}[\sqrt{-2}]$ by the remarks at the end of Section 7.5 and the fact that units of $\mathbb{Z}[\sqrt{-2}]$ are ± 1 (Exercise 7.4.5).

7.6.3. Suppose $x + \sqrt{-2} = (a + b\sqrt{-2})^3$ is a cube in $\mathbb{Z}[\sqrt{-2}]$, so a and b are ordinary integers. Deduce that

$$x = a^3 - 6ab^2 = a(a^2 - 6b^2) \quad \text{and} \quad 1 = 3a^2b - 2b^3 = b(3a^2 - 2b^2).$$

7.6.4. Deduce from Exercise 7.6.3 that $b = \pm 1$, $a = \pm 1$ and hence the only natural number solution for x is 5.

7.7* Factorizing a Sum of Two Squares

Diophantus' identity

$$(a^2 + b^2)(c^2 + d^2) = (ac - bd)^2 + (ad + bc)^2$$

tells us that when we multiply sums of two squares the product is also a sum of two squares. What happens when we divide? Is a divisor of a sum of two squares a sum of two squares? If a and b have a common divisor d, then $a^2 + b^2$ has the divisor d^2, which is trivially the sum $0^2 + d^2$ of two integer squares. And if we stick to a and b with no common prime divisor we have the following elegant theorem, discovered by Euler in 1747.

Divisors of sums of two squares. *If* $\gcd(a, b) = 1$, *then any divisor of* $a^2 + b^2$ *is of the form* $c^2 + d^2$, *where* $\gcd(c, d) = 1$.

Proof Each divisor $e > 1$ of $a^2 + b^2$ is a product of *Gaussian* prime divisors of $a^2 + b^2$, by unique prime factorization in $\mathbb{Z}[i]$. Because $a^2 + b^2 = (a + ib)(a - ib)$, each Gaussian prime divisor $q + ir$ of $a^2 + b^2$ divides either $a + ib$ or $a - ib$. And because $\gcd(a, b) = 1$, none of the divisors $q + ir$ is a real prime p, as $\frac{a}{p} \pm i\frac{b}{p}$ is not in $\mathbb{Z}[i]$.

Now the Gaussian prime divisors of e occur in conjugate pairs $q + ir$, $q - ir$, because if $q + ir$ divides e so does $q - ir$, by taking conjugates. From each pair we collect the member dividing $a + ib$, and form their product $c + id$. Then the conjugate members dividing $a - ib$ have product $c - id$, and

$$e = (c + id)(c - id) = c^2 + d^2.$$

Also $\gcd(c, d) = 1$, because a common (real) prime divisor p of c and d would divide $a + ib$ and hence both a and b, contrary to assumption. $\square$

This proof is another example of the way $\mathbb{Z}[i]$ refines our understanding of $\mathbb{Z}$. It shows that factorization into natural numbers of the form $x^2 + y^2$ can be viewed as a consequence of factorization into Gaussian integers. Moreover, it is *simpler* to view the situation this way, as the proof using real integers alone is more complicated.

Exercises

There is a similar theorem about divisors of numbers of the form $a^2 + 2b^2$, and it may be proved similarly using unique prime factorization in $\mathbb{Z}[\sqrt{-2}]$.

7.7.1. If $\gcd(a, b) = 1$, show that any divisor of $a + b\sqrt{-2}$ is of the form $c + d\sqrt{-2}$ with $\gcd(c, d) = 1$.

7.7.2. Deduce from Exercise 7.7.1 that if $\gcd(a, b) = 1$ then any divisor of $a^2 + 2b^2$ is of the form $c^2 + 2d^2$ with $\gcd(c, d) = 1$.

Euler's theorem on the divisors of $a^2 + b^2$ ties up nicely with the idea of "factorizing" Pythagorean triples explored in the exercises to Section 5.4.

7.7.3. Show that a divisor $c^2 + d^2$ of $a^2 + b^2$ corresponds to a Pythagorean triple $(2cd, b^2 - c^2, b^2 + c^2)$, which is a "factor" of the triple $(2ab, a^2 - b^2, a^2 + b^2)$. Illustrate this result with the triple $(319, 360, 481)$ from Plimpton 322.

7.8   Discussion

Complex Numbers and Geometry

The geometry of complex numbers is a vast subject. It covers not only the Euclidean plane but also the sphere, the non-Euclidean plane and even non-Euclidean space. On all of these objects, it is possible to describe isometries by simple functions of a complex variable. We have seen how this happens when the Euclidean plane is interpreted as $\mathbb{C}$, and it happens similarly on the sphere and the non-Euclidean plane when they are suitably mapped to the Euclidean plane.

A complex coordinate on the sphere is obtained by stereographic projection (Section 4.6*) from the sphere to the plane $\mathbb{C}$. This projects every point on the sphere, except the north pole, to a complex number z we take as its coordinate. We take ∞ as the coordinate of the north pole, which works perfectly in this situation. In particular, the half turn of the sphere about the real axis that exchanges the north and south poles sends the point with coordinate z to the point with coordinate $1/z$, so ∞ is exchanged with $1/\infty = 0$, as it should be. General rotations of the sphere turn out to be the functions of the form $\frac{az+b}{-\bar{b}z+\bar{a}}$.

The non-Euclidean plane mentioned in Section 3.9* has an obvious complex coordinate, because it is naturally viewed as the half plane of complex numbers $x + iy$ with $y > 0$. In fact, this is how it was introduced by Poincaré (1882), and he went on to show that its orientation-preserving isometries are the functions $\frac{az+b}{cz+d}$ with a, b, c, d real and $ad - bc > 0$. Examples are the function $2z$, which "translates" points z along the imaginary axis, and $-1/z$, which is a "half turn" about the point i. The simplest orientation-reversing isometry is reflection $-\bar{z}$ of z in the imaginary axis.

Figure 7.6 shows a tessellation of the half plane by triangles with angles $\pi/2$, $\pi/3$, and $\pi/7$. It is clear enough to the eye that all the triangles have the same angles, but they are also *congruent* in the sense of non-Euclidean geometry. The picture was in fact generated from one triangle by repeatedly reflecting in its sides. One of its symmetries is a translation along the imaginary axis.

Projection from line to circle (Section 4.6*) can be extended to a map $f(z) = \frac{z-i}{z+i}$ of the half plane onto the unit disk $\{z : |z| < 1\}$. (Incidentally, this accounts for the formula $\frac{t-i}{t+i}$ in Exercise 4.3.2 that gives all rational points on the circle as t runs through the rationals.) This correspondence between the half plane and the disk allows the latter to be used as another "model" of the non-Euclidean plane, much as one uses different map projections in geography to model the sphere. The half plane and disk models look somewhat similar, because the function $f(z) = \frac{z-i}{z+i}$ preserves angles and circles.

The disk model also reveals a striking algebraic analogy between the sphere and the non-Euclidean plane. When we use z as a coordinate in the disk model, its orientation-preserving isometries are

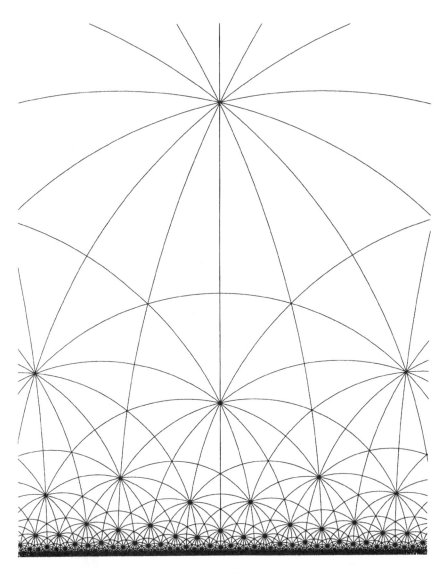

FIGURE 7.6 (2,3,7) tessellation of the half plane.

the functions $\frac{az+b}{\bar{b}z+\bar{a}}$, differing from those of the sphere only by one
− sign. This is just one of many ways in which the non-Euclidean
plane is "opposite" to the sphere. It has many parallel lines and the
sphere has none, its triangles have angle sum $< \pi$ and those of the
sphere have angle sum $> \pi$, and so on.

Poincaré (1883) generalized the half plane model to a "half space
model" of *non-Euclidean space* by considering the upper half of three-
dimensional space, the half above the (x, y)-plane say. He defined
the isometries of this space to be products of reflections in spheres
with centers on the (x, y)-plane. Reflection in a sphere is defined
analogously to reflection in a circle (Section 3.9*) and gives a geometry of the half space analogous to the non-Euclidean geometry of
the half plane.

Despite the third dimension, the isometries of non-Euclidean
space can be represented by functions of the complex variable $z =
x + iy$. This is because the spheres of reflection are determined by
the circles in which they cut the (x, y)-plane, so reflections in the
latter circles determine the reflections in the spheres above them.
Products of these reflections in circles then determine all isometries
of the half space, and in this way Poincaré found that the orientation-
preserving isometries of non-Euclidean space correspond to all the
complex functions $\frac{az+b}{cz+d}$ with $ad - bc \neq 0$.

Notice that in all cases the isometries are represented by *linear
fractional* functions of z, that is, quotients of linear functions $az + b$
and $cz + d$.

Poincaré's space explains why isometries are linear fractional
functions in all three geometries of surfaces—Euclidean plane,
sphere, and non-Euclidean plane—*these surfaces inherit their isometries from non-Euclidean space.* All three geometries actually occur
in non-Euclidean space: Euclidean planes as planes parallel to the
(x, y)-plane, spheres as ordinary spheres lying completely in the
upper half space, and non-Euclidean planes as vertical half planes
and hemispheres with their centers on the (x, y)-plane. This amazing
unification of geometry was discovered by Eugenio Beltrami (1868),
though it was Poincaré (1883) who first linked the geometries by
linear fractional functions. For more details, see Stillwell (1992) or
the papers of Beltrami and Poincaré in Stillwell (1996).

Quadratic Forms

The story of sums of squares, which started another thread in the history of complex numbers, also took a turn toward non-Euclidean geometry in the 19th century. To see how this came about, we have to say more about the work of Fermat and Lagrange.

Fermat's two squares theorem, describing the primes of the form $x^2 + y^2$, was the first of several such theorems. Fermat also described the primes of the form $x^2 + 2y^2$ (they are the primes $p \equiv 1$ or 3 (mod 8)) and the form $x^2 + 3y^2$ (they are the primes $p \equiv 1$ (mod 3)). However, he failed to find any such description of the primes of the form $x^2 + 5y^2$. The reasons for this did not become completely clear for another two centuries, but Lagrange made important progress in 1773.

Lagrange decided to develop a general theory of *binary quadratic forms*, that is, functions of the form $ax^2 + bxy + cy^2$, where a, b, c are integer constants and x, y are integer variables. Problems of Fermat's type then fall under the general problem of finding the possible values of a binary quadratic form, and in particular, finding the possible prime values. Lagrange noticed that many forms are *equivalent* in the sense that they are related by a change of variables. For example, if we substitute

$$x = x' + y',$$
$$y = y'$$

in $x^2 + y^2$ we get the form $x'^2 + 2x'y' + 2y'^2$, which takes exactly the same values. Why? Because as x' and y' run through all pairs of integers, so do $x = x' + y'$ and $y = y'$. Forms related by such a change of variables are called *equivalent* because they have the same sets of values.

It is not hard to work out that the transformations

$$x = ax' + by',$$
$$y = cx' + dy'$$

relating equivalent forms are those for which a, b, c, d are integers and $ad - bc = \pm 1$. Such transformations are now called *unimodular*. Lagrange used them to find the "simplest" form in a class of equivalent forms, and in this way found more efficient proofs of Fermat's theorems and many others.

The story turned geometric when Gauss noticed that the equivalents of a given form could be viewed as points in the upper half plane, related by functions

$$f(z) = \frac{az + b}{cz + d}$$

corresponding to the unimodular transformations. Because f is a linear fractional function with real coefficients, it is an isometry if the half plane is interpreted as the non-Euclidean plane. Gauss didn't realize what kind of geometry he was looking at here (though in fact he had speculated about non-Euclidean geometry in an abstract way), but Poincaré did, and he used geometric insights to help understand quadratic forms.

Poincaré realized, in fact, that the real problem was to understand unimodular transformations, which form a nonabelian group. This was the first nonabelian group encountered in number theory, and the first time the group concept was used to make a bridge from number theory to geometry, where the problem could be more easily understood.

Quadratic Integers and Lattices

In this chapter we have given several impressive results that follow from unique prime factorization in the rings of quadratic integers $\mathbb{Z}[i]$ and $\mathbb{Z}[\sqrt{-2}]$. Nevertheless, some readers may feel that unique prime factorization is a trivial property, which doesn't deserve the credit for Fermat's two squares theorem (say) or for showing that there is only one positive integer solution of $y^3 = x^2 + 2$.

In fact, unique prime factorization cannot be taken for granted, because it is sometimes *false*, and it is worth taking a closer look at the conditions that make it possible.

The proofs of unique prime factorization in $\mathbb{Z}[i]$ and $\mathbb{Z}[\sqrt{-2}]$ depend on finding a division property, like the one for $\mathbb{Z}$, which depends in turn on the "shape" of $\mathbb{Z}[i]$ and $\mathbb{Z}[\sqrt{-2}]$ in the plane $\mathbb{C}$. For example, to establish the division property of $\mathbb{Z}[i]$ we used the fact that its members lie at the corners of a square grid, and hence so do the Gaussian integer multiples of any number $\beta \neq 0$.

It is not clear that this process will always work, and it will certainly fail for $\mathbb{Z}[\sqrt{-5}]$, because $\mathbb{Z}[\sqrt{-5}]$ does *not* have unique prime factorization. An example that shows this is

$$6 = 2 \times 3 = (1 + \sqrt{-5})(1 - \sqrt{-5}).$$

The factors 2,3 and $1 + \sqrt{-5}$, $1 - \sqrt{-5}$ are all primes in $\mathbb{Z}[\sqrt{-5}]$, as can be seen most easily by using the norm

$$N(a + b\sqrt{-5}) = |a + b\sqrt{-5}|^2 = a^2 + 5b^2.$$

The norms of 2, 3, $1 + \sqrt{-5}$, $1 - \sqrt{-5}$ are 2^2, 3^2, 6, 6, respectively, and the proper divisors 2 and 3 of these norms are *not* norms of any numbers in $\mathbb{Z}[\sqrt{-5}]$. Hence 2, 3, $1 + \sqrt{-5}$, $1 - \sqrt{-5}$ have no proper divisors, and therefore they are primes.

The failure of unique prime factorization in $\mathbb{Z}[\sqrt{-5}]$ can also be explained by a geometric property, which neatly distinguishes $\mathbb{Z}[\sqrt{-5}]$ from $\mathbb{Z}[i]$ and $\mathbb{Z}[\sqrt{-2}]$. All of these rings are abelian groups under addition of complex numbers, and they and their subgroups are called *lattices* (because that is what they look like if their neighboring members are joined by lines; look again at Figure 7.4).

Prime factorization is related to certain subgroups called *ideals*. An ideal I in a ring R is a subgroup with the additional property that

(a member of I) $\times$ (a member of R) = (a member of I).

It turns out that each ideal in $\mathbb{Z}[i]$ is of a specially simple type called *principal*; it consists of all the multiples of some nonzero member β. The same is true of ideals in $\mathbb{Z}[\sqrt{-2}]$. In fact, this explains algebraically why prime factorization is unique in $\mathbb{Z}[i]$ and $\mathbb{Z}[\sqrt{-2}]$, because unique prime factorization is true of any ring in which all ideals are principal (the proof is basically the proof of the prime divisor property in Section 1.6). It follows that in a ring such as $\mathbb{Z}[\sqrt{-5}]$, where prime factorization is not unique, there will be ideals that are not principal, and hence not the same shape as the ring itself.

This allows us to "see" the failure of unique prime factorization in $\mathbb{Z}[\sqrt{-5}]$, in the shape of a nonprincipal ideal. Figure 7.7 shows $\mathbb{Z}[\sqrt{-5}]$, with stars marking the members of the ideal consisting of sums of multiples of 2 and $1 + \sqrt{-5}$. It is clear that the lattice of

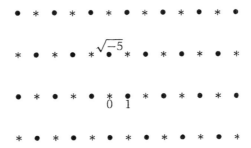

FIGURE 7.7   A nonprincipal ideal in $\mathbb{Z}[\sqrt{-5}]$.

stars is not rectangular, whereas $\mathbb{Z}[\sqrt{-5}]$ itself is. Hence the lattice is a nonprincipal ideal.

These examples give only a glimpse of the fascinating structure behind unique prime factorization in the quadratic integers. Readers are urged to consult Artin (1991) for details and Dedekind (1877) for the history of the subject. However, I cannot resist making one more tantalizing remark, because it unites the current train of thought with the previous one. Fermat's trouble with $x^2 + 5y^2$ is due to the failure of unique prime factorization in $\mathbb{Z}[\sqrt{-5}]$ and in fact *quadratic forms and quadratic integers are really the same subject*. Both depend on the study of lattice shapes, and lattice shapes are most naturally located in the non-Euclidean plane. Thus all roads lead to non-Euclidean geometry!

8

C H A P T E R

Conic Sections

8.1 Too Much, Too Little, and Just Right

Conic sections, as their name suggests, are curves obtained by cutting a cone by a plane. They have been studied since ancient times, originally because of their affinity with the circle, and with revived interest since the 17th century when it was found that they model the paths of projectiles, comets, and planets. Another motive for studying them is their ability to "construct" numbers not constructible by ruler and compass, such as $\sqrt[3]{2}$. Perhaps the best way to explain why the same curves arise in these apparently unrelated situations is to say that conic sections are the *simplest* curves, apart from straight lines. Therefore, of all the curves that can turn up in the world of mathematics, the conic sections will turn up most often.

Their simplicity is measured by the degree of their equations, something that was unknown to the ancients, but independently discovered by Fermat and Descartes when they invented analytic geometry. As we know, straight lines have equations of degree 1, $ax + by = c$. The conic sections (or *conics*, as they are often called) are the curves with equations of degree 2.

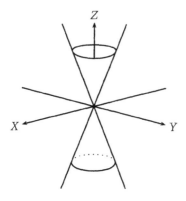

FIGURE 8.1   The cone.

To see why conics have degree 2, consider the cone $x^2 + y^2 = k^2 z^2$ shown in Figure 8.1. First of all, why is this the equation of a cone? One sees that the horizontal sections $z = $ constant are the circles $x^2 + y^2 = (k \times \text{constant})^2$. It follows that the surface $x^2 + y^2 = k^2 z^2$ is symmetric about the z-axis, and hence all sections by vertical planes through the z-axis must look the same. But the section through the plane $x = 0$ is $y^2 = k^2 z^2$, which is the pair of lines $y = \pm kz$, so the surface is in fact the cone obtained by rotating these lines about the z-axis.

The conic sections proper are the intersections of the cone with planes not passing through the origin. Such a plane can meet the cone in three different ways, and the corresponding curves of intersection (Figure 8.2) are called the *hyperbola*, *ellipse*, and *parabola*, from the Greek meaning roughly "too much," "too little," and "just right." Other English words with the same origin are "hyperbole" (something exaggerated or excessive), "ellipsis" (something cut short), and "parable" (something that runs alongside).

The same broad classification "too much," "too little," and "just right" occurs elsewhere in mathematics. For example, geometries and differential equations are both divided into hyperbolic, elliptic, and parabolic types. The parabolic case is always the exceptional, transitional case between hyperbolic and elliptic. Among the conic sections, the parabola is the exceptional case where the cutting plane is parallel to one of the lines in the cone. This happens when the cutting plane has slope $\pm k$. Hyperbolas occur when the cutting plane

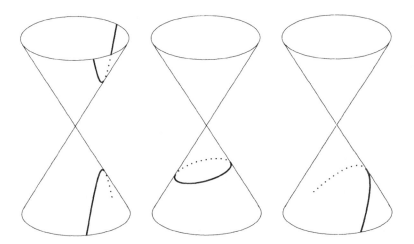

FIGURE 8.2   Hyperbola, ellipse, and parabola as sections of a cone.

slopes "too much" and cuts both halves of the cone, ellipses occur when the cutting plane slopes "too little," and cuts the cone in only a finite curve.

The geometric difference between hyperbola, ellipse, and parabola can also be recognized by algebra. Because the cone is symmetrical about the z-axis, there is no loss of generality in assuming the cutting plane to be perpendicular to the (y, z)-plane, so its equation is of the form $cy + dz = e$. In fact, dividing by the coefficient of z (or by the coefficient of y if the coefficient of z is zero), we get the equation of the cutting plane to be either

$$cy + z = e \qquad \text{or} \qquad y = e.$$

If the latter, we substitute $y = e$ in the equation to the cone and get

$$k^2 z^2 - x^2 = e^2$$

as the equation of the resulting hyperbola. If the former, we substitute $z = e - cy$ in the equation $x^2 + y^2 = k^2 z^2$ of the cone and get

$$x^2 + y^2(1 - k^2 c^2) + 2k^2 cey = k^2 e^2.$$

Thus the coefficient of y^2 depends on the slope $-c$ of the cutting plane. It is

- Less than 0 if $c^2 > k^2$, which happens if the conic section is a hyperbola.
- Greater than 0 if $c^2 < k^2$, which happens if the conic section is an ellipse.
- Equal to 0 if $c^2 = k^2$, which happens if the conic section is a parabola.

The equation $x^2 + y^2(1 - k^2c^2) + 2k^2cey = k^2e^2$ is the relation between x and y on the conic section, so it is really the equation of the *projection* of the curve in the (x, y)-plane. However, if we introduce coordinates in the cutting plane itself, the only change is to multiply the y-coordinate by a constant factor. It remains true that the coefficients of x^2 and y^2 have opposite signs for a hyperbola, they have the same sign for an ellipse, and there is no y^2 term for a parabola.

We can rewrite the equation

$$x^2 + y^2(1 - k^2c^2) + 2k^2cey = k^2e^2$$

in the form

$$x^2 = Dy + C^2 \qquad \text{or} \qquad x^2 + A(y^2 + 2By) = C^2$$

according as $1 - k^2c^2 = 0$ or not. Both these equations can be simplified by replacing y by y plus a suitable constant, that is, by a change of origin. The first becomes

$$x^2 = Dy \qquad \text{when } y \text{ is replaced by } y - C^2/D,$$

and the second becomes ("completing the square")

$$x^2 + Ay^2 = C^2 - AB^2 \qquad \text{when } y \text{ is replaced by } y - B.$$

With a little further tidying—dividing through to make the constant term 1, writing positive coefficients as squares and negative coefficients as negatives of squares—we finally obtain the simplest possible equations for the conic sections:

$$\frac{x^2}{a^2} - \frac{y^2}{b^2} = 1 \qquad \text{(hyperbola)}$$

$$\frac{x^2}{a^2} + \frac{y^2}{b^2} = 1 \qquad \text{(ellipse)}$$

$$y = ax^2 \qquad \text{(parabola)},$$

where a and b are nonzero constants.

Exercises

It is possible to check that all nonzero values of a and b actually arise from sections of cones, though this is a little tedious to do directly. In the case of the parabola, it is better to do the following.

8.1.1. Show that all parabolas have the same shape. In particular, if x, y are replaced by cx, cy for a suitable constant c, show that the equation $y = ax^2$ becomes $y = x^2$.

The shape of a hyperbola or ellipse is determined by the ratio of the coefficients of x^2 and y^2.

8.1.2. Show that hyperbolas of arbitrary shape occur as vertical sections of the cone $x^2 + y^2 = k^2 z^2$ as k varies.

8.1.3. Show that ellipses of arbitrary shape come from cutting the cone $x^2 + y^2 = z^2$ by suitable planes $y - dz = 1$.

Ellipses can also be obtained as sections of a circular cylinder.

8.1.4. Write down the equation of a circular cylinder symmetric about the z-axis, and find the equation of its intersection with the plane $y = mz$, using suitable coordinates in the latter.

8.2 Properties of Conic Sections

The conic sections have many interesting properties, and in this book we can mention only a few, as we wish to concentrate on our main theme, the relations between numbers, geometry, and functions. However, it is impossible to resist a brief look at some of the properties that make conic sections physically significant. These properties are not closely related to the cone; they come to light when the conic sections are seen from a different geometric viewpoint.

A conic section C can be defined in terms of a point F called its *focus* and a line $\mathcal{D}$ called its *directrix*. C is simply the set of points whose distance from F is a constant multiple of its perpendicular distance from $\mathcal{D}$ (Figure 8.3). The multiple is called the *eccentricity*, e.

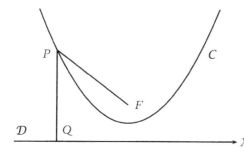

FIGURE 8.3   Focus and directrix of a conic section.

We take $\mathcal{D}$ to be the x-axis and F to be the point $(0, 1)$ for convenience. Then if $P = (x, y)$ we have

$$FP = \sqrt{x^2 + (y - 1)^2} \quad \text{and} \quad PQ = y,$$

and therefore

$$\sqrt{x^2 + (y - 1)^2} = ey.$$

Squaring both sides gives

$$x^2 + (y - 1)^2 = e^2 y^2,$$

and therefore

$$x^2 + (1 - e^2)y^2 - 2y + 1 = 0.$$

For $e = 1$ this is the parabola $2y = x^2 + 1$. For $e < 1$ it is an ellipse, and for $e > 1$ a hyperbola, as may be checked by completing the square and shifting the origin.

The focus is physically significant in Newton's theory of gravitation, because planets and comets travel on conic sections with the sun at their focus. (Of course, this is an idealization of the real situation. The *mathematical* situation, which closely approximates the real one, assumes two point masses with an inverse square law of attraction. Then if one mass is taken as the origin of coordinates, the other moves along a conic section with the origin as its focus.) This is a very famous result, but it would be a big detour for us to prove it. Instead, we shall prove another important property of the focus, which is the reason for its name, the Latin word for *fireplace*.[1] Kepler

---

[1] This meaning is also evident in the word for focus used in German, "Brennpunkt," meaning burning point.

gave it this name, knowing that if rays from a distant source fall directly onto a parabolic mirror, they are all reflected to the focus, and hence heat is concentrated there.

Focal property of the parabola.   *Lines parallel to the axis of symmetry of a parabola are reflected through the focus*

Proof   Given any point P on the parabola, consider the perpendicular from P to Q on the directrix (Figure 8.4). Then if F is the focus, the focus-directrix property gives

$$FP = PQ.$$

It follows that the equidistant line $\mathcal{T}$ of F and Q meets the parabola at P. We wish to show that $\mathcal{T}$ does not meet the parabola at any other point, so that $\mathcal{T}$ is the tangent at P.

If, on the contrary, $\mathcal{T}$ meets the parabola at a second point P', then F and Q are also equidistant from P', so $FP' = P'Q$. But the focus-directrix property of P' says that $FP' = P'Q'$, where $Q' \neq Q$ is the perpendicular projection of P' onto the directrix. This implies $P'Q = P'Q'$, which is contrary to Pythagoras' theorem.

Thus $\mathcal{T}$ is indeed the tangent at P. Because $\mathcal{T}$ is also the equidistant line of F and Q, the angles marked ϕ in Figure 8.4 are equal, and hence so are the angles marked θ, where $\mathcal{N}$ is the normal at P. In particular, this means that the vertical line striking the inside of the parabola at P is reflected through the focus F.   □

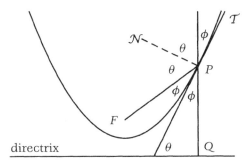

FIGURE 8.4   Tangent, normal, focus, and directrix of the parabola.

Exercises

The focal property of the parabola was first proved by the Greek mathematician Diocles, in a book *On Burning Mirrors* written around 200 B.C. As the title suggests, Diocles was aware of the potential applications of the theorem, and there is indeed a story that Archimedes used burning mirrors against Roman ships. It is probably only a legend, however; the Greeks did not care much for practical applications of geometry, and this one is of doubtful practicality in any case.

8.2.1. Comment on the feasibility of building a parabolic mirror to burn ships.

The most comprehensive ancient book on conic sections is the *Conics* of Apollonius. It was also written around 200 B.C., but it does not mention the focus of the parabola, although it includes a proof of a more difficult focal property of the ellipse.

To give a modern proof of this theorem, it helps to know that the ellipse $\frac{x^2}{a^2} + \frac{y^2}{b^2} = 1$ has eccentricity $e = \sqrt{1 - b^2/a^2}$, foci at $(\pm ae, 0)$, and directrices $x = \pm a/e$. (There are two of each because of the obvious symmetry of the ellipse.) Once this is known, it is relatively easy to check the focus-directrix property.

8.2.2. Check that the distance from $(ae, 0)$ to any point $P = (x, y)$ on the ellipse is e times the distance of P from the line $x = a/e$.

The focus-directrix property has a practical consequence known as the "thread construction of an ellipse."

8.2.3. Deduce from Exercise 8.2.2 that the sum of distances from the foci to any point on the ellipse is constant.

Thus if a length of thread is tied to a pair of nails at the foci F_1 and F_2 (Figure 8.5) and pulled tight by a pencil at P, then the pencil will draw an ellipse.

The focal property of the ellipse found by Apollonius states that the lines F_1P and F_2P make equal angles with the tangent at P. To prove this, let $\mathcal{T}$ be the line through P that *does* make equal angles with F_2P and F_1P. The problem then is to show that $\mathcal{T}$ does not meet the ellipse at a second point P', so that $\mathcal{T}$ is the tangent.

The latter problem is solved by a classic argument, which shows that the line F_1PF_2 reflected off the line $\mathcal{T}$ is the *shortest* path from F_1 to $\mathcal{T}$ to F_2.

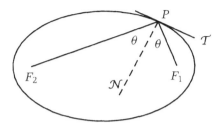

FIGURE 8.5 Foci, tangent, and normal of an ellipse.

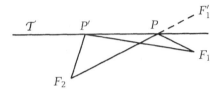

FIGURE 8.6 Minimality of the path of reflection.

8.2.4. By considering the reflection F_1' of F_1 in $\mathcal{T}$ (Figure 8.6) and the triangle inequality, show that F_1PF_2 is shorter than any other path $F_1P'F_2$ from F_1 to $\mathcal{T}$ to F_2.

8.2.5. Deduce from Exercise 8.2.4 and the constancy of the sum of focal distances in the ellipse that $\mathcal{T}$ is the tangent to the ellipse at P.

This proves the focal property of the ellipse. If we fix one focus of the ellipse and let the other tend to infinity, we obtain the parabola as an "ellipse with one focus at infinity." The lines through the focus at infinity become parallel, and thus the focal property of the parabola is a limiting case of the focal property of the ellipse.

8.2.6. Investigate whether there is an analogous focal property of the hyperbola.

8.3 Quadratic Curves

The calculations of Section 8.1 show that all conic sections are quadratic curves, because their equations take the form

$$\frac{x^2}{a^2} - \frac{y^2}{b^2} = 1 \qquad \text{(hyperbola)}$$

$$\frac{x^2}{a^2} + \frac{y^2}{b^2} = 1 \qquad\qquad \text{(ellipse)}$$

$$y = ax^2 \qquad\qquad \text{(parabola)}$$

when axes are suitably chosen. Thus in each case the equation can be written in the form $q(x, y) = 0$ where q is a quadratic polynomial.

However, it is not yet clear that all quadratic curves are conic sections. There is, in fact, a trivial exception—pairs of straight lines. The straight lines $x = 3y$ and $x = 4y$, for example, can be combined into the quadratic equation $(x - 3y)(x - 4y) = 0$, so the pair is technically a quadratic curve. We call this pair of straight lines a *degenerate* quadratic curve because it results from a genuine curve $(x - 3y)(x - 4y) = c$ as c shrivels to 0. Another degenerate quadratic curve is represented by the equation $x^2 + y^2 = 0$ for the single point $(0, 0)$. This curve results from degeneration of the circle $x^2 + y^2 = c^2$.

Degenerate quadratic curves $q(x, y) = 0$ can be spotted when $q(x, y)$ splits into linear factors (possibly with complex coefficients), so they are not a problem. But what about the genuine curve $xy = 1$? Can we transform its equation into one of the three by suitable choice of axes?

The answer is that all nondegenerate quadratic curves have equations of one of the three types, relative to suitable axes. Hence the genuine quadratic curves are all conic sections. This discovery of Fermat and Descartes was very important in the development of geometry, because it showed for the first time that the algebraic concept of degree is geometrically significant. Their result can be obtained in two steps: rotation of axes and shift of origin.

Suppose that we are given the most general quadratic curve

$$ax^2 + bxy + cy^2 + dx + ey + f = 0.$$

1. We first make the substitution

$$x = x' \cos\theta + y' \sin\theta$$
$$y = -x' \sin\theta + y' \cos\theta,$$

which amounts to choosing x'- and y'-axes at angle θ to the x- and y- axes, by the sin and cos addition formulas of Section 5.3. By suitable choice of θ, $ax^2 + bxy + cy^2$ takes the form $a'x'^2 + c'y'^2$, as the following calculation shows:

$$ax^2 + bxy + cy^2 = a(x'^2 \cos^2 \theta + 2x'y' \cos \theta \sin \theta + y'^2 \sin^2 \theta)$$
$$+ b(-x'^2 \cos \theta \sin \theta + x'y'(\cos^2 \theta - \sin^2 \theta)$$
$$+ y'^2 \cos \theta \sin \theta)$$
$$+ c(x'^2 \sin^2 \theta - 2x'y' \cos \theta \sin \theta + y'^2 \cos^2 \theta).$$

The coefficient of $x'y'$ is

$$a \sin 2\theta + b \cos 2\theta - c \sin 2\theta = (a - c) \sin 2\theta + b \cos 2\theta;$$

hence to make it zero we need

$$(c - a) \sin 2\theta = b \cos 2\theta.$$

If $c - a = 0$, we can satisfy this equation by choosing θ so that $\cos 2\theta = 0$. Otherwise we choose θ so that $\tan 2\theta = b/(c - a)$, which is always possible because $\tan 2\theta$ takes all real values.

We are assuming $b \neq 0$ (otherwise there is no need to rotate the axes), hence in both cases $\sin 2\theta \neq 0$. In the first case, this is because $\cos 2\theta = 0$, and in the second because $\tan 2\theta \neq 0$. We can use this fact to show that the coefficients of x'^2 and y'^2,

$$a' = a \cos^2 \theta - b \cos \theta \sin \theta + c \sin^2 \theta$$
$$\text{and} \quad c' = a \sin^2 \theta + b \cos \theta \sin \theta + c \cos^2 \theta,$$

are not both zero. If they are, then adding the equations $a' = 0$ and $c' = 0$ gives $a + c = 0$ or $c = -a$. But in this case, subtracting the equation $a' = 0$ from $c' = 0$ gives $0 = b \sin 2\theta$, contrary to the facts that $b \neq 0$ and $\sin 2\theta \neq 0$.

2. Relative to the new axes, the curve has an equation of the form

$$a'x'^2 + c'y'^2 + d'x' + e'y' + f' = 0,$$

with a' and c' not both zero. Now we shift the origin, substituting $x'' + A$ for x' and $y'' + C$ for y'. This gives the equation

$$a'(x''^2 + 2Ax'' + A^2) + c'(y''^2 + 2Cy'' + C^2) + d'(x'' + A) + e'(y'' + C) + f' = 0.$$

If $a' \neq 0$ we make the coefficient $2a'A + d'$ of x'' zero by choosing $A = -d'/2a'$. If $b' \neq 0$ we make the coefficient $2c'C + e'$ of y'' zero by choosing $C = -e'/2c'$. Because a' and c' are not both zero, this gives an equation of one of the forms:

$$a''x''^2 + c''y''^2 = f'',$$
$$\text{or} \quad a''x''^2 = e''y'' + f'',$$
$$\text{or} \quad c''y''^2 = d''x'' + f''.$$

A further shift of origin and renaming of variables and constants converts the latter two of these equations to the standard equation of the parabola,

$$y = ax^2.$$

The former equation becomes the standard equation of the hyperbola or the ellipse, according as a'' and c'' have opposite or equal signs—unless the constant $f'' = 0$, in which case it represents a degenerate quadratic curve.

Exercises

The examples of degenerate quadratic curves given above—the pair of lines $x = 3y$ and $x = 4y$ and the single point $(0,0)$—could be regarded as conic sections. After all, a plane can cut a cone in a pair of intersecting lines or a single point.

8.3.1. Show, however, that there is a degenerate quadratic curve that is *not* a section of a cone. Is it a section of a "degenerate cone"?

8.3.2. The lines $\frac{y}{b} = \pm\frac{x}{a}$ are called the *asymptotes* of the hyperbola $\frac{x^2}{a^2} - \frac{y^2}{b^2} = 1$. Describe how the hyperbola is related to its asymptotes, both in terms of degeneration and in terms of the cone.

Rotation of axes allows us to identify the quadratic curve $xy = 1$ as a hyperbola. Can you guess in advance how far the axes should be rotated?

8.3.3. Show that the curve $xy = 1$ is the same (after suitable rotation of axes and renaming of variables) as the hyperbola $x^2 - y^2 = 2$.

This raises the question: how might we tell in advance whether a quadratic curve is a hyperbola, ellipse, or parabola? The answer is certainly known after step 1. The curve is a parabola if one of a' or c' is zero, it is a hyperbola if they have opposite signs, and it is an ellipse if they have the same sign. Thus the problem is to detect these properties of a' and c' from properties of a, b, and c.

These properties are most easily brought to light using matrices, so for the rest of this exercise set we shall assume a basic knowledge of matrices and determinants. (The only facts we actually need are that the product of matrices corresponds to the composition of substitutions, and that the determinant of a product is the product of the determinants. A really tenacious reader may be able to prove these facts from first principles.)

The substitution

$$x = x' \cos \theta + y' \sin \theta$$
$$y = -x' \sin \theta + y' \cos \theta,$$

is written in matrix notation as

$$\begin{bmatrix} x \\ y \end{bmatrix} = \begin{bmatrix} \cos \theta & \sin \theta \\ -\sin \theta & \cos \theta \end{bmatrix} \begin{bmatrix} x' \\ y' \end{bmatrix}.$$

The usefulness of matrices here is due to the fact that we can also write the quadratic form $ax^2 + bxy + cy^2$ as a matrix product, namely,

$$ax^2 + bxy + cy^2 = \begin{bmatrix} x & y \end{bmatrix} \begin{bmatrix} a & b/2 \\ b/2 & c \end{bmatrix} \begin{bmatrix} x \\ y \end{bmatrix}.$$

8.3.4. Deduce that if $a'x'^2 + b'x'y' + c'y'^2 = \begin{bmatrix} x' & y' \end{bmatrix} \begin{bmatrix} a' & b'/2 \\ b'/2 & c' \end{bmatrix} \begin{bmatrix} x' \\ y' \end{bmatrix}$ is the quadratic form resulting from $ax^2 + bxy + cy^2$ by the substitution, then

$$\begin{bmatrix} a' & b'/2 \\ b'/2 & c' \end{bmatrix} = \begin{bmatrix} \cos \theta & -\sin \theta \\ \sin \theta & \cos \theta \end{bmatrix} \begin{bmatrix} a & b/2 \\ b/2 & c \end{bmatrix} \begin{bmatrix} \cos \theta & \sin \theta \\ -\sin \theta & \cos \theta \end{bmatrix}.$$

8.3.5. Deduce from Exercise 8.3.4, by taking determinants of both sides, that

$$b'^2 - 4a'c' = b^2 - 4ac.$$

8.3.6. Conclude from Exercise 8.3.5 that a nondegenerate quadratic curve

$$ax^2 + bxy + cy^2 + dx + ey + f = 0$$

is

- a hyperbola if $b^2 - 4ac > 0$,
- an ellipse if $b^2 - 4ac < 0$,
- a parabola if $b^2 - 4ac = 0$.

8.4* Intersections

As far as we know, the conic sections were first studied by Menaechmus, a Greek mathematician who lived around 350 B.C. Menaechmus was searching for a construction of $\sqrt[3]{2}$, which, as we now know (from Exercises 3.2.5* to 3.2.8*, for example), is not constructible by ruler and compass. This was not known in ancient Greece, but apparently it was suspected, because constructions of $\sqrt[3]{2}$ were sought using curves other than straight lines and circles. Menaechmus' solution was the simplest, because it used only quadratic curves, the hyperbola, and the parabola.

In terms of coordinates, his construction is almost a triviality. One takes the parabola $y = x^2/2$ and intersects it with the hyperbola $xy = 1$. At the intersection, $y = x^2/2 = 1/x$: therefore $x^3 = 2$ and so $x = \sqrt[3]{2}$. It is also not difficult to regard the parabola and hyperbola as "constructible" in a reasonable sense. We have seen the "thread construction" of the ellipse in Exercise 8.2.3, and there are many other mechanical constructions of conic sections. Assuming such constructions are available and that they allow the curves to be constructed from their coefficients, we can study numbers *constructible from conic sections*. As with ruler and compass constructions, the idea is to form intersections of conic sections and straight lines, then to use the resulting points to construct further lines and conics, and so on. Because the circle is a particular conic section, these constructions will include the ruler and compass constructions, and square roots in particular.

It is not hard to generalize Menaechmus' construction to show that if a number is constructible from conic sections, then so is its cube root. Thus *the numbers constructible from conic sections include all numbers obtainable from 1 by rational operations, square roots, and cube roots.*

The converse statement is also true. It depends on finding the equations that arise from intersections of conic sections, and solving these equations by rational operations, square roots, and cube roots. Again, I shall not do all the details, but I hope to explain why square roots and cube roots are crucial. (The details of this part can be completed by doing the exercises, however.)

As we know from the last two sections, conic sections are quadratic curves, and any conic may be brought into one of the standard forms

$$\frac{x^2}{a^2} - \frac{y^2}{b^2} = 1 \qquad \text{(hyperbola)}$$

$$\frac{x^2}{a^2} + \frac{y^2}{b^2} = 1 \qquad \text{(ellipse)}$$

$$y = ax^2 \qquad \text{(parabola)}$$

by rotation of axes and shift of origin. The amount of shift is determined by a rational computation, and the rotation depends on finding $\cos\theta$ and $\sin\theta$ when $\tan 2\theta$ is known. Because

$$\tan 2\theta = \frac{\sin 2\theta}{\cos 2\theta} = \frac{\sin 2\theta}{\sqrt{1 - \sin^2 2\theta}},$$

we can find $\sin 2\theta$ from $\tan 2\theta$ by solving a quadratic equation. Then because

$$\sin 2\theta = 2\sin\theta\cos\theta = 2\sin\theta\sqrt{1 - \sin^2\theta},$$

we can find $\sin\theta$ from $\sin 2\theta$ by solving another quadratic equation, and we can find $\cos\theta$ similarly. This means we can find the coefficients of the standard form conic by rational operations and square roots. It also means that the coefficients of *any* given conic, relative to the new axes, are computable by rational operations and square roots.

The good thing about the standard form equations is that they can all be solved for y as a function of x using rational operations and (at most one) square root. Namely:

$$y = b\sqrt{\frac{x^2}{a^2} - 1} \qquad \text{(hyperbola)},$$

$$y = b\sqrt{1 - \frac{x^2}{a^2}} \qquad\qquad \text{(ellipse)},$$

$$y = ax^2 \qquad\qquad \text{(parabola)}.$$

Now suppose we wish to find the intersection of the standard form conic C_1 with any other conic C_2, whose equation relative to the new axes is, say,

$$Ax^2 + Bxy + Cy^2 + Dx + Ey + F = 0. \qquad\qquad (C_2)$$

If C_1 is $y = ax^2$, we substitute ax^2 for y in C_2, obtaining a *quartic* (fourth-degree) equation for y. As will be seen in Exercises 8.4.1 and 8.4.2, this equation can be solved by rational operations, square roots, and cube roots.

If C_1 is $y = b\sqrt{\pm\left(1 - \frac{x^2}{a^2}\right)}$, we substitute this for y in C_2, obtaining

$$Ax^2 + Bbx\sqrt{\pm\left(1 - \frac{x^2}{a^2}\right)} \pm Cb^2\left(1 - \frac{x^2}{a^2}\right) + Dx + Eb\sqrt{\pm\left(1 - \frac{x^2}{a^2}\right)} + F = 0$$

or

$$Ax^2 \pm Cb^2\left(1 - \frac{x^2}{a^2}\right) + Dx + F = b(-Bx - E)\sqrt{\pm\left(1 - \frac{x^2}{a^2}\right)}.$$

Squaring both sides of this gives a quartic equation for x, hence it again follows that it can be solved by rational operations, square roots, and cube roots.

To sum up, what we have shown in outline is the following: *the numbers constructible by conic sections are precisely those obtainable from 1 by rational operations, square roots, and cube roots.*

Exercises

The solution of the quartic equation was discovered by Cardano's student Lodovico Ferrari in 1545. The first step is a small simplification of the equation by change of variable.

8.4.1. Show the general quartic equation $x^4 + ax^3 + bx^2 + cx + d = 0$ takes the form $x^4 + px^2 + qx + r = 0$ when x is replaced by $x + s$ for a suitable constant s.

The latter quartic equation can be rewritten

$$(x^2 + p)^2 = px^2 - qx + p^2 - r.$$

(Why?) This suggests that both sides might be made into squares simultaneously by adding an appropriate quantity. If so, we can reduce the quartic to a quadratic by taking the square root of both sides.

8.4.2. Show that the previous equation implies

$$(x^2 + p + y)^2 = (p + 2y)x^2 - qx + (p^2 - r + 2py + y^2)$$

for any y.

The left-hand side is a square by construction. The right-hand side is a quadratic in x, and we aim to make it a square by finding a suitable y.

8.4.3. Show that $Ax^2 + Bx + C = A(x + B/2A)^2$ when $B^2 - 4AC = 0$.

8.4.4. When $Ax^2 + Bx + C = (p + 2y)x^2 - qx + (p^2 - r + 2py + y^2)$, check that $B^2 - 4AC$ is a cubic in y. Hence conclude from the previous exercises, and Exercises 7.2.1 and 7.2.2, that the general quartic equation may be solved by rational operations, square roots, and cube roots.

8.5 Integer Points on Conics

Because conic sections are quadratic curves, we understand in principle all the rational points on a conic section C. Assuming the equation of C has rational coefficients, Section 4.4 gives the following description of all its rational points: they consist of any single rational point P on C, together with the other points where C is met by lines through P with rational slope. The rational points include the integer points, of course, but describing the integer points alone is another story entirely. The hyperbola, ellipse, and parabola all require different methods, with the hyperbola being the most difficult and interesting. To keep the story as simple as possible, I shall stick to equations in standard form.

Finding the integer points on the ellipse $\frac{x^2}{a^2} + \frac{y^2}{b^2} = 1$ is easiest, at least in principle. All its points lie within distance $\max(|a|, |b|)$ of the origin, so if necessary one can test all integer points within this radius to see whether they satisfy the equation.

The parabola $y = ax^2$ has infinitely many integer points, which are easily computed from a, assuming a is rational. We write a as a fraction in lowest terms, m/n, so the equation becomes $ny = mx^2$. Then each integer x makes mx^2, and hence ny, an integer, and so it gives an integer point just in case n divides mx^2. This in fact happens precisely for the multiples of the least positive x divisible by n.

The hyperbola is interesting even in the special case $x^2 - dy^2 = 1$, where d is a positive integer. In fact, we shall first study $x^2 - 2y^2 = 1$, and concentrate on its "positive branch," for which $x > 0$. There is one obvious integer point on this hyperbola, namely, $(1, 0)$, and the next is found by trial to be $(3, 2)$.

The point $(3, 2)$ is a "seed" that produces all the integer points on $x^2 - 2y^2 = 1$. Here we shall generate infinitely many integer points from it, and in the next chapter we'll show that they are *all* the integer points on the positive branch. The process of generation is surprising, because it uses the irrational number $\sqrt{2}$. We use $\sqrt{2}$ to define a "product" of integer points on $x^2 - 2y^2 = 1$ as follows.

Generation of integer points on $x^2 - 2y^2 = 1$**.** *If (m_1, n_1) and (m_2, n_2) are integer points on the hyperbola $x^2 - 2y^2 = 1$, then so is the point (m_3, n_3), where m_3 and n_3 are defined by*

$$m_3 + n_3\sqrt{2} = (m_1 + n_1\sqrt{2})(m_2 + n_2\sqrt{2}).$$

Proof   First we should make sure that m_3 and n_3 really are *defined* by

$$m_3 + n_3\sqrt{2} = (m_1 + n_1\sqrt{2})(m_2 + n_2\sqrt{2}).$$

The intention is to expand the right-hand side as

$$m_1 m_2 + 2n_1 n_2 + \sqrt{2}(n_1 m_2 + m_1 n_2)$$

and set $m_3 = m_1 m_2 + 2n_1 n_2$ and $n_3 = n_1 m_2 + m_1 n_2$ by "equating rational and irrational parts." But why is this valid? The reason is that $r + s\sqrt{2} = u + v\sqrt{2}$ for rational r, s, u, v only if $r = u$ and $s = v$;

if not, we have $\sqrt{2} = (r - u)/(v - s)$, contrary to the irrationality of $\sqrt{2}$.

It is clear from the definition of m_3 and n_3 that they are integers. Now because (m_1, n_1) and (m_2, n_2) are points on $x^2 - 2y^2 = 1$, we have

$$m_1^2 - 2n_1^2 = 1 \qquad \text{and} \qquad m_2^2 - 2n_2^2 = 1.$$

Factorizing these equations we get

$$(m_1 + n_1\sqrt{2})(m_1 - n_1\sqrt{2}) = 1 \quad \text{and} \quad (m_2 + n_2\sqrt{2})(m_2 - n_2\sqrt{2}) = 1,$$

and taking their product,

$$(m_1 + n_1\sqrt{2})(m_1 - n_1\sqrt{2})(m_2 + n_2\sqrt{2})(m_2 - n_2\sqrt{2}) = 1.$$

Rearranging the factors gives

$$(m_1 + n_1\sqrt{2})(m_2 + n_2\sqrt{2})(m_1 - n_1\sqrt{2})(m_2 - n_2\sqrt{2}) = 1,$$

which is in fact

$$(m_3 + n_3\sqrt{2})(m_3 - n_3\sqrt{2}) = 1.$$

The first factor comes from the definition of m_3 and n_3, and the second because changing $+$ to $-$ signs in the definition still gives a valid identity. Expanding the last equation, we get

$$m_3^2 - 2n_3^2 = 1,$$

so (m_3, n_3) is an integer point on $x^2 - 2y^2 = 1$, as required. $\square$

It follows from this result that the points (m_k, n_k) defined by

$$m_k + n_k\sqrt{2} = (3 + 2\sqrt{2})^k$$

are infinitely many integer points on the hyperbola $x^2 - 2y^2 = 1$. The result also has an obvious generalization for any nonsquare positive integer d, using the fact that $\sqrt{d}$ is irrational for such a d.

Generation of integer points on $x^2 - dy^2 = 1$. *If (m_1, n_1) and (m_2, n_2) are integer points on the hyperbola $x^2 - dy^2 = 1$, then so is the point (m_3, n_3), where m_3 and n_3 are defined by*

$$m_3 + n_3\sqrt{d} = (m_1 + n_1\sqrt{d})(m_2 + n_2\sqrt{d}).$$ $\square$

Exercises

In case you are wondering why we started with the hyperbola $x^2 - 2y^2 = 1$ instead of $x^2 - y^2 = 1$ ....

8.5.1. Observe that $x^2 - y^2 = (x + y)(x - y)$, and hence show that $(\pm 1, 0)$ are the only integer points on $x^2 - y^2 = 1$. What can you say about integer points on $x^2 - dy^2 = 1$ when d is an integer square?

The integer points (m_k, n_k) on $x^2 - 2y^2 = 1$ were known to the Greeks under the name of "side and diagonal numbers," because the ratios n_k/m_k approximate the ratio $\sqrt{2}$ between the diagonal and side of the square.

8.5.2. Check that the first few values of (m_k, n_k) are $(3, 2)$, $(17, 12)$, $(99, 70)$, and show that $n_k/m_k \to \sqrt{2}$ as $k \to \infty$. Give a geometric interpretation of this fact in terms of the asymptote $x = \sqrt{2}y$ of the hyperbola $x^2 - 2y^2 = 1$.

The pairs (m_k, n_k) are actually not all the side and diagonal number pairs. The Greeks discovered the sequence

$$(1, 1), \quad (3, 2), \quad (7, 5), \quad (17, 12), \quad (41, 29), \quad (99, 70), \quad \ldots$$

of pairs (x_k, y_k) that alternately satisfy $x^2 - 2y^2 = -1$ and $x^2 - 2y^2 = 1$. We are not sure how they discovered the sequence, but they computed it from $(x_1, y_1) = (1, 1)$ and the following *recurrence relations* giving (x_{k+1}, y_{k+1}) in terms of (x_k, y_k):

$$x_{k+1} = x_k + 2y_k,$$
$$y_{k+1} = x_k + y_k.$$

8.5.3. Show by induction that the sequence (x_k, y_k) defined by

$$x_k + y_k\sqrt{2} = (1 + \sqrt{2})^k$$

satisfies the recurrence relations and hence agrees with the sequence of pairs of side and diagonal numbers.

8.5.4. Deduce from Exercise 8.5.3 that $(m_k, n_k) = (x_{2k}, y_{2k})$.

8.5.5. Show also that the pairs (x_{2k+1}, y_{2k+1}) satisfy the equation $x^2 - 2y^2 = -1$.

It should be noted that the *negative* integer powers of $3 + 2\sqrt{2}$ also give integer points on $x^2 - 2y^2 = 1$, for the following reason.

8.5.6. Show that if $m_k + n_k\sqrt{2} = (3+2\sqrt{2})^k$ then $m_k - n_k\sqrt{2} = (3-2\sqrt{2})^k = (3+2\sqrt{2})^{-k}$.

Notice also that $(3 + 2\sqrt{2})^0 = 1 + 0\sqrt{2}$ gives the point $(1,0)$ on $x^2 - 2y^2 = 1$. Thus the points we have found so far correspond to members of the infinite cyclic group of numbers $(3+2\sqrt{2})^k$ for integers k. (See Section 6.10 for the definitions of abelian and cyclic groups.) There is in fact a way to treat the whole positive branch of the curve as a group, as we shall see in Chapter 9, and it then becomes clear that the points we have found so far are the subgroup of all integer points.

The deduction of $m_3^2 - 2n_3^2 = 1$ from $m_1^2 - 2n_1^2 = 1$ and $m_2^2 - 2n_2^2 = 1$ hints at another instance of a multiplicative norm, like the norm $N(a+bi)$ on $\mathbb{Z}[i]$ we studied in Section 7.4. Here we are dealing with the ring

$$\mathbb{Z}[\sqrt{2}] = \{a + b\sqrt{2} : a, b \in \mathbb{Z}\}$$

with the norm defined by

$$N(a + b\sqrt{2}) = a^2 - 2b^2.$$

We are interested in the members $a + b\sqrt{2}$ with norm 1, because they correspond to integer points on $x^2 - 2y^2 = 1$, but the norm is in fact multiplicative for all members of $\mathbb{Z}[\sqrt{2}]$.

8.5.7. Show that $N((a_1 + b_1\sqrt{2})(a_2 + b_2\sqrt{2})) = N(a_1 + b_1\sqrt{2})N(a_2 + b_2\sqrt{2})$.

8.5.8. Suggest a norm for $\mathbb{Z}[\sqrt{d}]$ and show that it is multiplicative.

(*Hint*: It may help to recall Brahmagupta's identity from the exercises to Section 7.1.)

8.6* Square Roots and the Euclidean Algorithm

The irrationality of $\sqrt{2}$ and other numbers tormented Greek mathematicians for hundreds of years and provoked many attempts to relate irrationals to integers in a comprehensible way. The most interesting, as far as square roots are concerned, is a generalization of the Euclidean algorithm. As Euclid himself described it, the Euclidean algorithm "continually subtracts the lesser number from the greater." Such a process can also be applied to a pair of numbers whose ratio is irrational, such as $\sqrt{2}$ and 1. Of course the algorithm will not terminate in this case, but if there is some pattern to the numbers produced it surely gives some new understanding of the nature of $\sqrt{2}$. And for $\sqrt{2}$ we get a pattern that is the next best thing to termination, namely, *periodicity*.

The pattern can be seen most easily by applying the Euclidean algorithm to the pair $(\sqrt{2} + 1, 1)$. The lesser number 1 can be subtracted twice from the greater, $\sqrt{2}+1$, producing the pair $(1, \sqrt{2}-1)$. It so happens that the new lesser number $\sqrt{2} - 1$ can also be subtracted twice from the new greater number 1, and the same thing happens again and again; it appears that the lesser number can *always* be subtracted twice from the greater number. But how can we be sure?

The fog clears miraculously if we view each number pair as adjacent sides of a rectangle and subtract the lesser number from the greater by cutting off the square on the lesser side. In particular, the first two subtractions are interpreted as cutting off unit squares from the rectangle with sides $\sqrt{2} + 1$ and 1, as shown in Figure 8.7.

This produces a rectangle with sides 1 and $\sqrt{2} - 1$ and the new rectangle is the same shape as the original. We can confirm this by computing the ratio of the sides:

$$\frac{1}{\sqrt{2}-1} = \frac{\sqrt{2}+1}{(\sqrt{2}-1)(\sqrt{2}+1)} = \frac{\sqrt{2}+1}{1}.$$

It follows that subtracting the lesser side twice from the greater will produce rectangles of the same shape indefinitely. Thus the Euclidean algorithm is *periodic* on the pair $(\sqrt{2}+1, 1)$ in the sense that it continually subtracts the lesser number twice from the greater.

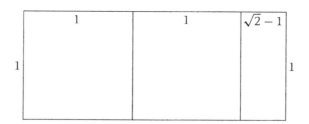

FIGURE 8.7   Periodicity and $\sqrt{2}$.

This means that if we use the algorithm in division-with-remainder form, the quotient at each step is 2.

As for the pair $(\sqrt{2}, 1)$, at first the lesser number is subtracted once from the greater, producing the pair $(1, \sqrt{2} - 1)$. But this is the pair we have just seen, so after the first subtraction the lesser number is always subtracted twice from the greater. Equivalently, the sequence of quotients in the division-with-remainder algorithm is $1, 2, 2, 2, 2, 2, \ldots$. We say that the Euclidean algorithm is *ultimately periodic* on the pair $(\sqrt{2}, 1)$.

The geometric explanation of the periodicity of the Euclidean algorithm also shows that each new pair is obtained from the previous pair by multiplying its members by $\sqrt{2} - 1$. Hence the $(k+1)$th pair is $(\sqrt{2} - 1)^k(\sqrt{2} + 1, 1)$. This links the periodicity of $\sqrt{2}$ with the process used in the previous section to generate integer points on the hyperbola $x^2 - 2y^2 = 1$.

Perhaps the most attractive way to display the periodicity of $\sqrt{2}$ is to work out its *continued fraction*. Recycling some of the facts found previously, we find

$$\sqrt{2} + 1 = 2 + \sqrt{2} - 1$$

$$= 2 + \cfrac{1}{\sqrt{2} + 1} \qquad \text{as we already know}$$

$$= 2 + \cfrac{1}{2 + \sqrt{2} - 1} \qquad \text{by the first line}$$

$$= 2 + \cfrac{1}{2 + \cfrac{1}{\sqrt{2}+1}} \qquad \text{by the second line}$$

$$= 2 + \cfrac{1}{2 + \cfrac{1}{2+\frac{1}{\sqrt{2}+1}}} \qquad \text{similarly,}$$

and so on. The limit of these fractions exists and is called the *continued fraction for* $\sqrt{2} + 1$. We write it

$$\sqrt{2} + 1 = 2 + \cfrac{1}{2 + \cfrac{1}{2+\cfrac{1}{2+\frac{1}{\ddots}}}},$$

and subtracting 1 from both sides gives the continued fraction for $\sqrt{2}$:

$$\sqrt{2} = 1 + \cfrac{1}{2 + \cfrac{1}{2 + \cfrac{1}{2 + \cfrac{1}{\ddots}}}}.$$

The sequence $1, 2, 2, 2, 2, 2, 2, 2, \ldots$ of natural numbers to the left of the + signs is the sequence of quotients occurring in the running of the Euclidean algorithm on $(\sqrt{2}, 1)$.

Exercises

The Euclidean algorithm on $(\sqrt{2}, 1)$ is linked not only with the integer points on the hyperbola $x^2 - 2y^2 = 1$ but also with the side and diagonal numbers mentioned in the previous exercise set. The Greeks may very well have discovered the side and diagonal numbers as the coefficients of $\sqrt{2}$ and 1 occurring in successive terms produced by the Euclidean algorithm.

8.6.1. Show that $(\sqrt{2} - 1)^k = (-1)^k(y_k\sqrt{2} - x_k)$, where y_k and x_k are the side and diagonal numbers defined in the previous exercise set.

8.6.2. Deduce from Exercise 8.6.1 the coefficients of each term produced from $(\sqrt{2}, 1)$ by the Euclidean algorithm form a side and diagonal number pair.

There is a similar relationship between $\sqrt{3}$, the Euclidean algorithm and integer points on $x^2 - 3y^2 = 1$. Again we find that $\sqrt{3} + 1$ has slightly simpler behavior than $\sqrt{3}$, by considering a rectangle of width $\sqrt{3} + 1$ and height 1.

8.6.3. Show that the large and small rectangles in Figure 8.8 are the same shape.

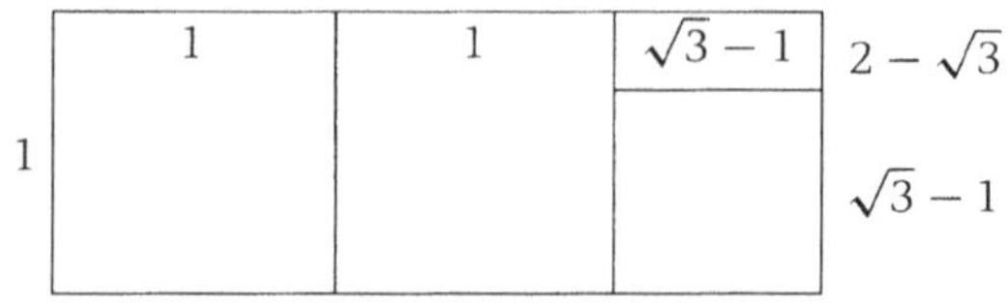

FIGURE 8.8   Periodicity and $\sqrt{3}$.

8.6.4. Deduce from Exercise 8.6.3 that

- the Euclidean algorithm is periodic on $(\sqrt{3} + 1, 1)$, with successive quotients $2, 1, 2, 1, 2, 1, 2, 1, \ldots,$

- $\sqrt{3} + 1 = 2 + \cfrac{1}{1 + \cfrac{1}{2 + \cfrac{1}{1 + \cfrac{1}{2 + \cfrac{1}{1 + \cfrac{1}{\ddots}}}}}}$,

- each period of the Euclidean algorithm on $(\sqrt{3} + 1, 1)$ reduces the size of the pair by a factor $2 - \sqrt{3}$.

Now define integers a_k, b_k by $(2 + \sqrt{3})^k = a_k + b_k\sqrt{3}$, or equivalently by $(2 - \sqrt{3})^k = a_k - b_k\sqrt{3}$.

8.6.5. Show that all the points (a_k, b_k) lie on the hyperbola $x^2 - 3y^2 = 1$.

As in the previous exercise set, these results can be interpreted as finding quadratic integers of norm 1 as powers of a "seed" quadratic integer of norm 1.

8.6.6. Interpret the previous result in terms of the norm $N(a + b\sqrt{3}) = a^2 - 3b^2$ on $\mathbb{Z}[\sqrt{3}]$, observing that $2 + \sqrt{3}$ and $2 - \sqrt{3}$ have norm 1.

8.7*   Pell's Equation

The equation $x^2 - dy^2 = 1$, where d is a nonsquare integer, is known as *Pell's equation*. John Pell was a 17th-century mathematician who had little or nothing to do with the equation, but Euler attached his name to it by mistake, and it stuck. The equation would be better named after Brahmagupta or Fermat, who solved it for particular values of d, or after Lagrange, who first gave the complete solution. We already know solutions for $d = 2$ and $d = 3$, where small solutions can be found by trial, and we have seen how these small solutions generate infinitely many others. For larger values of d, however, it is hard to find even one solution, apart from the trivial one $x = 1$, $y = 0$. The smallest nontrivial solution appears to vary wildly with d, and can be alarmingly large. Brahmagupta said "whoever can solve $x^2 - 92y^2 = 1$ in less than a year is a mathematician," and the equation $x^2 - 61y^2 = 1$ (posed by Fermat) is tougher still.

The smallest nontrivial solution of $x^2 - 92x^2 = 1$ is $x = 1151$, $y = 120$, while the smallest nontrivial solution of $x^2 - 61y^2 = 1$ is $x = 1766319049$, $y = 226153980$.

In such a situation, proving the *existence* of a nontrivial solution is easier than finding it. We shall in fact find infinitely many candidates for the smallest nontrivial solution, and show that one of them must be correct. The method of proof was invented by Dirichlet, and he called it the *pigeonhole principle*. The finite form of the principle says that if $n + 1$ pigeons are in n boxes then at least one box contains two pigeons. The infinite form of the principle says that if infinitely many pigeons are in finitely many boxes, then at least one box contains infinitely many pigeons. Both forms of the pigeonhole priciple are involved in the proof; we shall use the finite one first.

Dirichlet's approximation theorem.   *For any real number α and any integer $Q > 1$ there are integers p, q with $0 < q < Q$ and $|q\alpha - p| \leq 1/Q$.*

Proof   Consider the $Q + 1$ numbers

$$0, \quad 1, \quad \alpha - p_1, \quad 2\alpha - p_2, \quad \ldots, \quad (Q-1)\alpha - p_{Q-1},$$

where $p_1, p_2, \ldots, p_{Q-1}$ are integers chosen so that all the numbers lie in the interval from 0 to 1. If we divide this interval into subintervals of length $1/Q$, then we have Q subintervals containing $Q + 1$ numbers. Hence at least two numbers are in the same subinterval; that is, they are distance $\leq 1/Q$ apart. Because the difference between any two of the numbers is of the form $q\alpha - p$, for integers p and q with $0 < q < Q$, this means $|q\alpha - p| \leq 1/Q$ as required.   □

This theorem says that $q\alpha - p$ can be made at least as small as $1/q$, by suitable choice of p and q. It is particularly useful when α is irrational, because $q\alpha - p$ is never zero in that case, and hence we get infinitely many numbers $q\alpha - p$, each no larger than the corresponding $1/q$.

Here is how Dirichlet used his approximation theorem to show there are integers x and y such that $x^2 - dy^2 = 1$. The strategy is to make $p - q\sqrt{d}$ small enough that

$$p^2 - dq^2 = (p - q\sqrt{d})(p + q\sqrt{d}) \leq 3\sqrt{d}.$$

Thanks to the irrationality of $\sqrt{d}$, this gives infinitely many integers p and q for which $p^2 - dq^2 \leq 3\sqrt{d}$, and the infinite pigeonhole principle can then be used to show that some of them give $p^2 - dq^2 = 1$.

The other tool in the proof is the norm $N(p - q\sqrt{d}) = p^2 - dq^2$ and its multiplicative property, which can be verified by multiplying out both sides: $N((p_1 - q_1\sqrt{d})(p_2 - q_2\sqrt{d})) = N(p_1 - q_1\sqrt{d})N(p_2 - q_2\sqrt{d})$.

Existence of nontrivial solutions of $x^2 - dy^2 = 1$**.** *If $\sqrt{d}$ is irrational there are positive integers x and y such that $x^2 - dy^2 = 1$.*

Proof Applying the Dirichlet approximation theorem to $\alpha = \sqrt{d}$, for any integer $Q > 1$ we have positive integers p, q with

$$|p - q\sqrt{d}| \leq 1/Q \quad \text{and} \quad 0 < q < Q.$$

Because $\sqrt{d}$ is irrational, $p - q\sqrt{d} \neq 0$. By letting $Q \to \infty$ we therefore get infinitely many pairs of positive integers p, q with $|p - q\sqrt{d}| \leq 1/q$.

Now for any such pair

$$|p + q\sqrt{d}| \leq |p - q\sqrt{d} + 2q\sqrt{d}| \leq |p - q\sqrt{d}| + |2q\sqrt{d}| \leq 3q\sqrt{d},$$

hence

$$|p^2 - q^2 d| = |p + q\sqrt{d}||p - q\sqrt{d}| \leq 3q\sqrt{d}/q = 3\sqrt{d}.$$

Thus we have infinitely many pairs of positive integers p, q with $N(p - q\sqrt{d}) \leq 3\sqrt{d}$.

We now apply the infinite pigeonhole principle to obtain infinitely many pairs p, q with even more special properties.

1. Because there are only finitely many natural numbers $\leq 3\sqrt{d}$, infinitely many of the numbers $p - q\sqrt{d}$ have the same norm, say N.

2. Because there are only finitely many congruence classes mod N, infinitely many of the numbers $p - q\sqrt{d}$ with norm N have p in the same congruence class, and infinitely many of the latter numbers have q in the same congruence class.

To sum up, there is an infinite set of numbers $p - q\sqrt{d}$ with the same norm $N(p - q\sqrt{d}) = N$, all p in the same congruence class mod N, and all q in the same congruence class mod N.

Now take two numbers $p_1 - q_1\sqrt{d}$, $p_2 - q_2\sqrt{d}$ from this set and consider their quotient $x + y\sqrt{d}$, which has norm 1 by the multiplicative property of norm. I claim that x and y are integers. Because

$$\frac{p_1 - q_1\sqrt{d}}{p_2 - q_2\sqrt{d}} = \frac{(p_1 - q_1\sqrt{d})(p_2 + q_2\sqrt{d})}{p_2^2 - q_2^2 d}$$

$$= \frac{p_1 p_2 - q_1 q_2 d}{N} + \frac{p_1 q_2 - q_1 p_2}{N}\sqrt{d},$$

we have to prove N divides $p_1 p_2 - q_1 q_2 d$ and $p_1 q_2 - q_1 p_2$. By hypothesis,

$$p_1 \equiv p_2 \ (\text{mod } N) \quad \text{and} \quad q_1 \equiv q_2 \ (\text{mod } N),$$

hence

$$p_1 p_2 - q_1 q_2 d \equiv p_1^2 - q_1^2 d \equiv 0 \quad (\text{mod } N),$$

because

$$p_1^2 - q_1^2 d = N(p_1 - q_1 d) = N.$$

Thus N divides $p_1 p_2 - q_1 q_2 d$. It also follows, by multiplying the congruences $p_1 \equiv p_2 \ (\text{mod } N)$ and $q_2 \equiv q_1 \ (\text{mod } N)$, that $p_1 q_2 \equiv q_1 p_2$ $(\text{mod } N)$, and hence

$$p_1 q_2 - q_1 p_2 \equiv 0 \quad (\text{mod } N).$$

Thus N divides $p_1 q_2 - q_1 p_2$. This proves the claim that

$$\frac{p_1 - q_1\sqrt{d}}{p_2 - q_2\sqrt{d}} = x + y\sqrt{d} \quad \text{for some integers } x \text{ and } y,$$

and $1 = N(x + y\sqrt{d}) = x^2 - dy^2$. Finally, $y \neq 0$ because $p_1 - q_1\sqrt{d} \neq p_2 - q_2\sqrt{d}$, so this is a nontrivial solution. □

Exercises

Dirichlet's approximation theorem says there are rational numbers p/q "very close" to any irrational number α. For each $q > 1$ there are fewer than q^2 rationals with denominator $\leq q$ between successive integers, nevertheless ....

8.7.1. Deduce from Dirichlet's approximation theorem that there are infinitely many values of q for which there is a rational p/q at distance no more than $1/q^2$ from α.

Two instances of this close approximation phenomenon are the approximation $22/7$ to π and the even more remarkable approximation $355/113$ discovered by the Chinese mathematician Zǔ Chōngzhī (429–500 A.D.).

8.7.2. Using the numerical value $\pi = 3.14159265\ldots$, show that $22/7$ approximates π within $1/7^2$ and $355/113$ approximates π within $1/113^2$ (in fact much more closely).

The existence of a nontrivial solution to $x^2 - dy^2 = 1$ is connected with the periodicity of the continued fraction for $\sqrt{d}$, as one would imagine from the examples $d = 2$ and $d = 3$ studied in the previous section and its exercises. In fact the traditional method for finding a nontrivial solution of $x^2 - dy^2 = 1$ was to derive it from the ultimate periodicity of the continued fraction for $\sqrt{d}$. However, the periodicity result is somewhat harder, and for proofs we refer the reader to Stark (1978) or Baker (1984).

8.8 Discussion

The Projective View of Conic Sections

An interesting alternative to the process of cutting a cone by a plane is the process of *projecting a circle*. These two processes are much the same, but the concept of projection is worth a closer look, because it brings new ideas to the fore and actually leads to a whole new branch of mathematics: *projective geometry*. To grasp the idea of projection, imagine a vertical pane of glass with a circle C drawn on it, illuminated by light from a point P (Figure 8.9).

The shadow $\mathcal{K}$ of C on a horizontal plane is a conic section:

- an ellipse if P is above the top of C,

- a parabola if P is level with the top of C,

- a hyperbola if P is below the top of C (but above the bottom of C).

So far, this is just a way to produce a cone (the cone of rays through P and C) and cut it by a plane (the horizontal plane), and hence obtain a conic section (the shadow of C). But it gets more interesting when

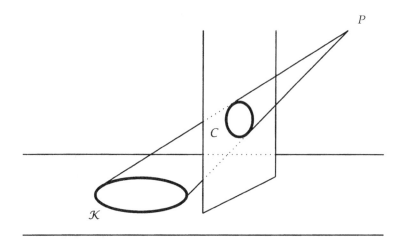

FIGURE 8.9   A conic section as the projection of a circle.

the light rays are reversed. Imagine that your eye is at the point P, receiving light rays emitted by a conic section $\mathcal{K}$ in the horizontal plane. The view of $\mathcal{K}$ seen when looking straight ahead can be captured by drawing, on a vertical window, a curve C that appears to cover $\mathcal{K}$. This in fact is the method used by Renaissance artists to draw three-dimensional scenes in correct perspective. The window was called "Alberti's veil," and Figure 8.10 is a woodcut by Albrecht Dürer showing how to use it.

Because any conic section is the projection of a circle, it follows that *any conic section looks like a circle, when suitably viewed.* But in that case, how do we identify which kind of conic section $\mathcal{K}$ we are viewing from a point P? We do so by observing the position of the circle, C, relative to the *horizon*, which is:

- above C when $\mathcal{K}$ is an ellipse, because in this case P is above C,

- tangential to the top of C when $\mathcal{K}$ is a parabola, because in this case P is level with the top of C,

- through two points of C when $\mathcal{K}$ is a hyperbola, because in this case P is below the top of C.

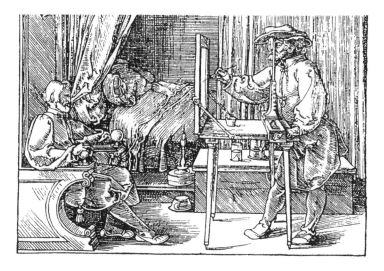

FIGURE 8.10   Dürer woodcut showing how to use Alberti's veil.

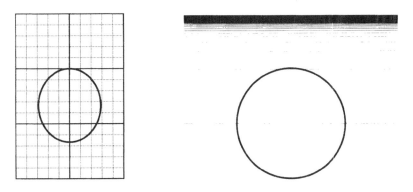

FIGURE 8.11   Ellipse and its perspective view.

Each figure shows two views of the conic section in question. Looking down on the horizontal plane, they appear as they normally do; looking toward the horizon, each looks like a circle, because of the peculiar choice of viewpoint P. The views toward the horizon also include parallel lines in the horizontal plane to establish where the

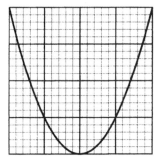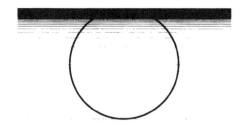

FIGURE 8.12   Parabola and its perspective view.

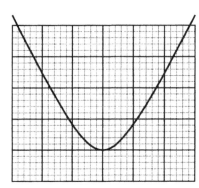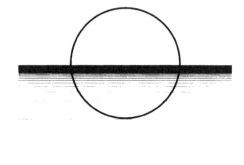

FIGURE 8.13   Hyperbola and its perspective view.

horizon is. These lines are actually equally spaced in the horizontal plane, but of course they appear to "accumulate" at the horizon.

(In the view of the hyperbola, the portion above the horizon corresponds to the other branch of the hyperbola, behind the eye. A real eye, of course, can't see this, but we can imagine it seen by a "mathematically ideal" eye, which can see in all directions.)

The interesting thing about the horizon is that it represents points that *do not belong* to the horizontal plane. These points are aptly called *points at infinity* because they are where actual points appear to go as they move far away. The horizon itself is called the *line at infinity*, and the horizontal plane together with its line at infinity is called the *projective plane*. *Projective geometry* is the study of the projective plane and the transformations of it (such as projection) that map lines to lines. This geometry is much less discriminating than Euclidean geometry, because it thinks every conic section looks like a circle, but it can distinguish between them very simply once the line at infinity is given:

- an ellipse is a conic section with no points at infinity,
- a parabola is a conic section with one point at infinity,
- a hyperbola is a conic section with two points at infinity.

This classification, which goes back to Kepler, gives another sense in which the hyperbola, ellipse, and parabola are "too much", "too little," and "just right." There is also a neat connection with the other discovery of Kepler (and Newton), that a celestial body travels on a conic section under the influence of the sun's gravity. The conic is a hyperbola if the body has more than enough energy to go to infinity, an ellipse if it has too little energy, and a parabola if the energy is just right.

9 Elementary Functions

9.1 Algebraic and Transcendental Functions

The main theme of this book has been the tension between arithmetic and geometry and its creative role in the development of mathematics. The story of $\sqrt{2}$ is an excellent example of such tension and its beneficial effects: geometry confronted arithmetic with the diagonal of the unit square, arithmetic expanded its concept of number in response, and the new number $\sqrt{2}$ proved its worth by giving new insight into the old numbers, for example, by generating integer solutions of the equation $x^2 - 2y^2 = 1$ (Section 8.5). In other cases, geometry was not so much a source of conflict with arithmetic as a source of immediate insight; for example, in generating Pythagorean triples by the chord construction (Section 4.3) or in guaranteeing unique prime factorization in the Gaussian integers by the triangle inequality (Section 7.5).

From the other side, arithmetic confronted geometry with the problem of describing curves in terms of numbers, addition, and multiplication. Geometry responded with coordinates and polynomial equations, which were a brilliant success with conic sections and many other curves. This development went hand in hand with another expansion of arithmetic—into the algebra of polynomials and rational functions.

The most complete fusion of arithmetic and geometry was obtained with the concept of *algebraic curve* —a curve defined by a polynomial equation $p(x, y) = 0$ in the two variables x and y—and the corresponding concept of an *algebraic function* y of x. (The symmetry of the definition shows that the inverse function x of y is also algebraic. Strictly speaking, neither may be a true function—for example, more than one value of y may correspond to the same value of x—but this is not important here.) Is algebra then the perfect reconciliation of arithmetic and geometry?

Not quite. Geometry has more challenges to offer, such as the concept of angle and the related sine and cosine functions. The function $\sin x$ is *not* an algebraic function of x and, equivalently, the curve $y = \sin x$ is not an algebraic curve. The reason, already seen in the exercises to Section 5.2, is that $y = \sin x$ does not contain the x-axis, yet it meets the x-axis at infinitely many points. No algebraic curve has this property, because its intersections with the x-axis satisfy the polynomial equation $p(x, 0) = 0$, which has only finitely many solutions.

Thus $y = \sin x$ is not an algebraic curve and $\sin x$ is not an algebraic function. Functions that are not algebraic are called *transcendental*, because they "transcend" algebra. Thinking back to the definition of cos and sin, one recalls that they were defined as functions of arc length of the unit circle, and *arc length* was defined as the least upper bound of the lengths of polygons. The least upper bound exists, by the completeness of the real numbers, but it would be nice to have a more explicit description of its value. We can now see that an algebraic description, at any rate, is out of the question. The arc length of the circle cannot be an algebraic function (of the coordinates of its endpoints, say), otherwise the sine function would also be algebraic.

Geometry creates many transcendental functions. Any nonconstant periodic function, such as the arc length of a closed curve, must be transcendental for the same reason as the sine; its graph meets a straight line in infinitely many points. In fact, the arc length function of almost any algebraic curve is transcendental, and very often the area function of the curve is also transcendental. Such functions cannot be fully investigated by algebra, and eventually one turns to calculus for further enlightenment. But there is no need to rush into advanced methods. The most common transcendental functions arise from conic sections and inherit many of their properties from elementary geometry and algebra. For example, the addition formulas for cos and sin are inherited from the geometry of the circle, as we saw in Section 5.3.

In the next two sections we shall see that the most important transcendental function, the exponential, inherits its basic properties from the geometry of the hyperbola. The exponential function can be defined as a function of area, and the area under the hyperbola has properties that are clear from geometry. Yet the exponential function also reveals properties of the hyperbola that are not otherwise obvious. In particular, we shall see that it highlights the integer points in a remarkable way, so once again there is an unexpected rapprochement between arithmetic and geometry.

Exercises

Arc length and area happen to be essentially the same function for the unit circle, because the area of a sector with angle θ is $\theta/2$. This can be seen by approximating the area by triangles with unit sides and angle θ/n.

9.1.1. Show that the area of an isosceles triangle with two unit sides and angle θ/n between them is $\frac{1}{2} \sin \frac{\theta}{n}$.

We have not yet given a general definition of area for curved figures, but the area of a sector is naturally interpreted as a limit of polygon areas.

9.1.2. Show, using Exercise 5.2.7, that $n \times \frac{1}{2} \sin \frac{\theta}{n} \to \frac{\theta}{2}$ as $n \to \infty$, and explain why this limit can be interpreted as the area of the unit sector with angle θ.

This result shows that the parameter θ in the equations $x = \cos\theta$, $y = \sin\theta$ can also be interpreted as (twice) an area. For other curves, the area and arc length functions are not so closely related, and area usually turns out to be more manageable. This is certainly true for the hyperbola, and it is the reason we use area to define the transcendental functions used in the remainder of this chapter.

9.2   The Area Bounded by a Curve

The idea of approximating a curve by polygons can be used to define the area of a curved region, just as easily as we define the length of a curve. The case that interests us is where the curve is the graph $y = f(x)$ of an algebraic function f, and the region is bounded by the curve, the x-axis, and the vertical lines $x = a$ and $x = b$ (Figure 9.1).

The area of the region is the least upper bound of the areas of all polygons Π contained in it. It exists if there is an upper bound to these polygon areas, as there is, for example, if there is an upper bound to $f(x)$ itself over the interval from a to b.

Students of calculus will recognize this definition as essentially the definite integral of $f(x)$ from a to b, but we shall not need calculus to derive the basic properties of area. For example, the following property of curved areas is inherited from the corresponding property of polygon areas: *if a region is magnified by a factor M in the x-direction and a factor N in the y-direction, then its area is magnified by a factor MN.* Magnification by MN is true for the areas of polygons Π (for example, by cutting them into triangles with bases in

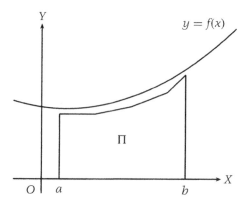

FIGURE 9.1   Area under a curve.

the x-direction and heights in the y-direction), and hence it is true for their least upper bound.

An important special case of this result is that magnification by M in all directions magnifies area by M^2. This is the underlying reason why the area of a circle of radius R is proportional to R^2.

Exercises

The first known determination of a curved area was made by Hippocrates of Chios around 430 B.C. He found the area of the region between the two circular arcs shown in Figure 9.2: one a quarter circle with radius OB, the other a semicircle with diameter AB. The region is called a *lune* because of its resemblance to a crescent moon, and Hippocrates showed that it has the same area as triangle AOB. Approximation by polygons is not required, except to show that the area of a circle is proportional to the square of its radius. (Hippocrates probably just assumed this; the idea of proving it rigorously using approximation by polygons is credited to Eudoxus, around 350 B.C.).

9.2.1. Show that the quarter circle with radius OB has the same area as the semicircle with diameter AB, and deduce Hippocrates' result.

Hippocrates made history by showing that curved areas are not beyond the grasp of mathematics, but his actual result is not very informative. It says nothing about the area of the circle, because the areas of the semicircle and quarter circle cancel out. Indeed, because we now

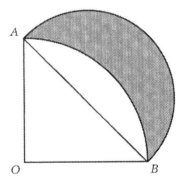

FIGURE 9.2   The lune of Hippocrates.

know that the area function of the circle is transcendental, an algebraic area (like that of triangle AOB) can only be obtained by subtracting one transcendental area from another.

The simplest curve whose area function happens to be algebraic is the parabola, and this area function was also the first to be discovered, by Archimedes around 250 B.C. A modern proof of his result may be based on approximation by polygons like those shown in Figure 9.3.

For simplicity, we take the parabola to be $y = x^2$ and find the area it bounds with the x-axis and the line $x = 1$. The proof is easily generalized to find the area up to an arbitrary line $x = a$.

9.2.2. Show that the area of a polygon with steps of width $1/n$ as in Figure 9.3 is $(1^2 + 2^2 + 3^2 + \cdots + (n-1)^2)/n^3$. We shall call this the nth *lower step polygon.*

This raises the problem of summing the series $1^2 + 2^2 + 3^2 + \cdots + (n-1)^2$, which in fact was solved by Archimedes for another purpose. (Strangely enough, he used this series to find the area of a spiral.)

9.2.3. Show by induction on n that

$$1^2 + 2^2 + 3^2 + \cdots + (n-1)^2 = \frac{2n^3 - 3n^2 + n}{6}.$$

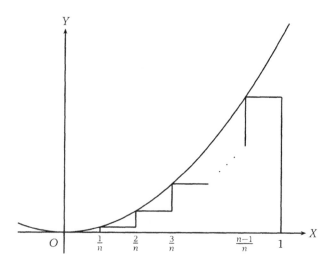

FIGURE 9.3 Area under a parabola.

9.2.4. Deduce from Exercise 9.2.3 that the area of the nth lower step polygon $\to 1/3$ as $n \to \infty$.

It looks like we have found the area under the curve between 0 and 1, but to be sure we should check that $\frac{1}{3}$ is really the least upper bound of the areas of all polygons in this region.

9.2.5. Find an nth *upper* step polygon that contains the region and differs from the nth lower step polygon by area $\frac{1}{n}$. Conclude that $\frac{1}{3}$ is the only number between the areas of lower and upper step polygons, and hence that it is the least upper bound of the areas of all polygons in the region below the curve between 0 and 1.

9.3 The Natural Logarithm and the Exponential

The algebraic area function for the parabola is another instance where the parabola is "just right." We already know that the circle has a transcendental area function, and the same is true of other ellipses. If we attempt to find the area under the hyperbola $xy = 1$, between $x = 1$ and $x = t$, say, we are in for another disappointment. The resulting function of t is also transcendental, so we cannot expect to "see" the least upper bound of polygonal areas, as we could for the parabola. But while the area function itself is complicated, some of its *properties* are simple because they are inherited from the hyperbola.

The area under $xy = 1$ from $x = 1$ to $x = t$ is called the *natural logarithm* of t and is written $\log t$. It follows that $\log 1 = 0$, and it is natural to suppose $\log t$ is negative for $t < 1$. We ensure this by taking the area with a negative sign when $0 < t < 1$. Figure 9.4 shows the graph of $\log t$ for $t > 0$. For the moment we do not attempt to define log for other values of t.

The most important property of the logarithm is the following, which follows very directly from the fact that $y = 1/x$.

Additive property of log. *If a and b are positive real numbers, then*

$$\log ab = \log a + \log b.$$

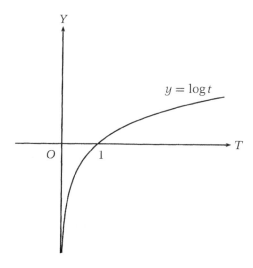

FIGURE 9.4 Graph of the logarithm function.

Proof We consider the area under the curve $xy = 1$ between $x = 1$ and $x = ab$, which is $\log ab$ by definition, and we split it by the line $x = a$ (Figure 9.5). This cuts off a region of area $\log a$, again by definition, so it remains to show that the region between $x = a$ and $x = ab$ has area $\log b$.

If we compare the region between a and ab with the region between 1 and b whose area is $\log b$ by definition (Figure 9.6), we see that the former region is a times as long as the latter. But it is also $1/a$ times as high. In fact, because $y = 1/x$, the height at the point $x = at$ between a and ab is $1/at$, which is $1/a$ times the height $1/t$ at the corresponding point $x = t$ between 1 and b.

Thus the region between a and ab is the result of magnifying a region of area $\log b$ by a factor a in the x-direction and a factor $1/a$ in the y-direction; hence its area is also $\log b$, as required.

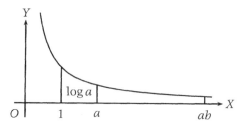

FIGURE 9.5 The area defining $\log a$.

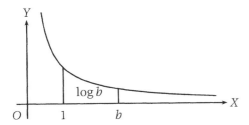

FIGURE 9.6 The area defining $\log b$.

This establishes the additive property when $a, b \geq 1$, but the argument is similar when either of them is between 0 and 1. One finds, for example, that the region from 1 to ab is smaller than the region from 1 to a when $a > 1$ and $b < 1$, and this is accounted for because the area from 1 to b has a negative sign. □

It follows from the additive property of logarithms that $\log t$ behaves like an *exponent* of t, that is, as if $t = e^{\log t}$ for some number e. The number e is the value of t for which $\log t = 1$, and its basic properties (including its existence, which is not completely obvious) are included in the following theorem.

Consequences of the additive property. *The* log *function has the properties*

1. *For any real number a and integer n, $\log a^n = n \log a$.*

2. *The* log *function takes each real value exactly once.*

3. *There is a number e with $\log e = 1$.*

4. *Suppose a^r is defined, for any real number r, to be the number whose* log *is $r \log a$. Then $x = e^y$ if and only if $y = \log x$.*

Proof

1. When $n = 2, 3, \ldots$ the additive property gives

$$\log a^2 = \log a + \log a = 2 \log a,$$
$$\log a^3 = \log a^2 + \log a = 2 \log a + \log a = 3 \log a,$$

and so on (or more formally, by induction on n). When $n = 0$,

$$\log a^0 = \log 1 = 0,$$

from the definition of log as an area, and when $n = -1$ we have

$$0 = \log 1 = \log aa^{-1} = \log a + \log a^{-1},$$

which implies

$$\log a^{-1} = -\log a.$$

Finally,

$$\log a^{-2} = \log a^{-1} + \log a^{-1} = -2 \log a,$$

and so on, by the additive property again.

2. If $a > 1$ then $\log a > 0$ from the definition of log as an area, and because $\log a^n = n \log a$ we can get arbitrarily large values of the log function by choosing sufficiently large values of the integer n. We can also get *arbitrarily closely spaced* values of the log function, by first choosing a near 1 so as to make $\log a$ as small as we please, then taking the equally spaced values $\log a^2$, $\log a^3$, ... It follows, by an argument like that for the completeness of the real numbers in Section 3.4, that for any real number ρ we can separate the values of $\log t$ into a lower set whose least upper bound is ρ and an upper set whose greatest lower bound is ρ.

Now it is clear from its definition that $\log t$ increases with t, hence the values of t are separated into a lower set L, for which $\log t$ has least upper bound ρ and an upper set U for which $\log t$ has greatest lower bound ρ. But the only number between the two sets of values of $\log t$ is ρ, hence it follows from the strict increase of the log function that *if τ is the least upper bound of L then $\log \tau$ must be ρ.*

3. It follows that there is a number e such that $\log e = 1$.

4. More generally, any real number $r \log a$ is a value of $\log t$. We define a^r to be the number whose log is $r \log a$, because this is consistent with the meaning of a^n for integers n by part 1. It then follows that e^y is the number x whose log is $y \log e = y$. That is, $x = e^y$ if and only if $y = \log x$. □

The fourth property is also described by saying that the *exponential function e^y* of y is the *inverse* of the log function. Thus the graph of $t = e^y$ (Figure 9.7) results from the graph of $y = \log t$ (Figure 9.4) by swapping the t- and y-axes. A consequence of the inverse relationship is that $t = e^{\log t}$, as claimed earlier.

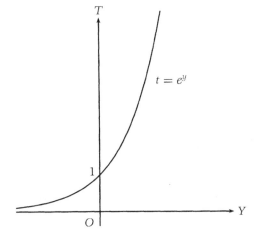

$t = e^y$

1

O → Y

FIGURE 9.7   Graph of the exponential function.

Exercises

In the previous exercise set we saw that the area under the parabola is related to $(1^2+2^2+3^2+\cdots+n^2)/n^3$, which we were able to evaluate exactly. The area under the hyperbola is similarly related to $1 + \frac{1}{2} + \frac{1}{3} + \cdots + \frac{1}{n}$, which we do not understand as well. At first it is not even clear whether this sum has a limit as $n \to \infty$, but a quick way to find out is to use what we now know about the area under the hyperbola.

9.3.1. Representing $1 + \frac{1}{2} + \frac{1}{3} + \cdots + \frac{1}{n}$ by a suitable step polygon, show that it is greater than $\log n$, and hence that $1 + \frac{1}{2} + \frac{1}{3} + \cdots + \frac{1}{n} \to \infty$ as $n \to \infty$.

In fact, the "rate of growth" of $1 + \frac{1}{2} + \frac{1}{3} + \cdots + \frac{1}{n}$ is amazingly close to that of $\log n$. Euler discovered that $1 + \frac{1}{2} + \frac{1}{3} + \cdots + \frac{1}{n} - \log n$ tends to a constant of value approximately 0.5772 as $n \to \infty$. The constant is known as *Euler's constant*, and it has been computed to many decimal places, but it is not known whether it is rational or irrational.

9.3.2. By using upper and lower step polygons, show that

$$1 + \frac{1}{2} + \frac{1}{3} + \cdots + \frac{1}{n-1} > \log n > \frac{1}{2} + \frac{1}{3} + \cdots + \frac{1}{n},$$

and hence deduce that $0 < 1 + \frac{1}{2} + \frac{1}{3} + \cdots + \frac{1}{n} - \log n < 1$.

9.3.3. Show by area considerations that $1 + \frac{1}{2} + \frac{1}{3} + \cdots + \frac{1}{n-1} - \log n$ is increasing and hence has a limit as $n \to \infty$ by the completeness of the real numbers. Show that $1 + \frac{1}{2} + \frac{1}{3} + \cdots + \frac{1}{n} - \log n$ has the same limit.

9.4   The Exponential Function

The additive property of the log function translates into the following property of the exponential function, which one recognizes as the characteristic property of exponents in general.

Addition formula for the exponential function.   *For real a and b,*

$$e^{a+b} = e^a e^b.$$

Proof   By definition of the exponential function, e^{a+b} is the number whose log is $a + b$. By the additive property of log, it follows that e^{a+b} is the product of the numbers whose logs are a and b. That is, $e^{a+b} = e^a e^b$.   □

Thus e^x behaves like the xth power of e, and we are justified in using a notation that suggests this. But why use powers of the mysterious number e instead of powers of something familiar, like 2 or 10? Most of the reasons for the convenience of e can be traced back to the fact that $\log e = 1$. The properties of general exponential functions a^x often involve a factor $\log a$ (see the exercises), and hence they are simplest when $\log a = 1$.

Other formulas showing e^x to be the simplest exponential function are

$$e^x = \lim_{n \to \infty} \left(1 + \frac{x}{n}\right)^n,$$

$$e^x = 1 + \frac{x}{1!} + \frac{x^2}{2!} + \frac{x^3}{3!} + \cdots.$$

We shall not need these formulas, but the second one, in particular, is connected with a spectacular discovery of Euler (1748) we cannot fail to mention:

$$e^{i\theta} = \cos\theta + i\sin\theta.$$

This can be proved by comparing

$$e^{i\theta} = 1 + \frac{i\theta}{1!} + \frac{(i\theta)^2}{2!} + \frac{(i\theta)^3}{3!} + \cdots$$

with known infinite series for $\cos\theta$ and $\sin\theta$:

$$\cos\theta = 1 - \frac{\theta^2}{2!} + \frac{\theta^4}{4!} - \frac{\theta^6}{6!} + \cdots,$$

$$\sin\theta = \theta - \frac{\theta^3}{3!} + \frac{\theta^5}{5!} - \frac{\theta^7}{7!} + \cdots.$$

Euler's formula is surely the most conclusive argument for the naturalness of e, but if anyone is not yet satisfied, consider the special case $\theta = \pi$:

$$e^{i\pi} = -1.$$

Only e can singlehandedly bring i and π down to earth!

Euler's formula incidentally proves that e^x, and hence its inverse $\log x$, is a transcendental function, as claimed earlier. If e^x were algebraic, then the equation $e^x = 1$ would have only finitely many solutions, real or imaginary. But in fact this equation has infinitely many solutions: $x = 2in\pi$ for all integers n.

Another thing we learn from Euler's formula is that the exponential function e^x is best regarded as a function of a *complex* variable x. It then embraces both the functions cos and sin, and their properties follow from properties of the exponential function. For example, the somewhat complicated addition formulas for cos and sin become consequences of the simple addition formula for e^x (as was verified for cis $\theta = e^{i\theta}$ in the exercises to Section 5.3). When cos and sin are subordinated to the exponential function in this way, one also understands their uncanny similarity to the so-called *hyperbolic cosine* and *hyperbolic sine*, cosh and sinh, which will be introduced in the next section.

Exercises

Because a^x is the number whose log is $x \log a$, namely, $e^{x \log a}$, powers of any positive real number a can be expressed as powers of e. The inverse of the power function a^x is called the *logarithm to base a* (so the ordinary log is logarithm to base e). If $y = a^x$ we write $x = \log_a y$.

9.4.1. Deduce from these definitions that $(e^a)^b = e^{ab}$, $e^x = a^{x/\log a}$, and $\log_a x = \log x / \log a$.

When we allow complex values of x, the exponential function takes the same value for many values of x, as noticed earlier. Hence its inverse, the log, is no longer a function. The single real value of $\log x$ for each real x is joined by infinitely many complex values. The happier side of this situation is that we can now find values of $\log x$ for *negative* real values of x, because any real $x \neq 0$ occurs as a value of the exponential function.

9.4.2. Find a value of $\log(-1)$.

To find all the values of $\log x$, it is necessary to know that

$$e^{a+ib} = e^a(\cos b + i \sin b) \quad \text{for any real } a \text{ and } b,$$

which follows from the addition formula and Euler's formula.

9.4.3. Assuming $e^{a+ib} = e^a(\cos b + i \sin b)$, find all values of $\log(-1)$.

We can also define a^b as $e^{b \log a}$ for complex numbers a and b, though the expression acquires infinitely many values from the infinitely many values of the log. An interesting example is i^i, which was first evaluated by Euler.

9.4.4. Use Euler's formula to show that $i = e^{i\pi/2}$.

9.4.5. Deduce from Exercises 9.4.1 and 9.4.4 that $e^{-\pi/2}$ is the real value of i^i. What are its other values?

Assuming the infinite series for e^x given earlier, it follows that

$$e = 1 + \frac{1}{1!} + \frac{1}{2!} + \frac{1}{3!} + \cdots .$$

The terms of this series tend to 0 very rapidly, which makes it easy to find approximate values of e, such as 2.718. The series also shows, however, that e is an irrational number.

9.4.6. Suppose on the contrary that $e = m/n$ for some integers m and $n \neq 0$. Show the following in turn:

- If $e = m/n$ then $n!e$ is an integer.
- On the other hand,

  $n!e = $ an integer $+ \frac{1}{n+1} + \frac{1}{(n+1)(n+2)} + \frac{1}{(n+1)(n+2)(n+3)} + \cdots$.
- $\frac{1}{n+1} + \frac{1}{(n+1)(n+2)} + \frac{1}{(n+1)(n+2)(n+3)} + \cdots < \frac{1}{2} + \frac{1}{2^2} + \frac{1}{2^3} + \cdots = 1$.
- This is a contradiction.

In fact, Charles Hermite showed in 1873 that e is a transcendental number. That is, e is not the root of any polynomial equation with integer coefficients. It was the first "known" number found to be transcendental. Building on Hermite's proof, Ferdinand Lindemann showed in 1882 that π is also transcendental. The proofs use calculus heavily and are a lot more difficult than the proofs that the exponential and circular functions are transcendental. Apparently numbers are harder to understand than functions, at least as far as transcendance goes.

9.5   The Hyperbolic Functions

Just as the circle $x^2 + y^2 = 1$ can be defined by the pair of functions $x = \cos\theta$ and $y = \sin\theta$, the hyperbola $x^2 - y^2 = 1$ can be defined by a pair of functions $x = \cosh\theta$ and $y = \sinh\theta$. It is even possible to interpret the parameter θ as (twice) the area of a "sector" of the hyperbola, and to define the functions cosh and sinh thereby, but to save time we define them so that the equation $\cosh^2\theta - \sinh^2\theta = 1$ is obvious.

The *hyperbolic cosine* $\cosh\theta$ and the *hyperbolic sine* $\sinh\theta$ are defined by

$$\cosh\theta = \frac{e^\theta + e^{-\theta}}{2},$$

$$\sinh\theta = \frac{e^\theta - e^{-\theta}}{2}.$$

It follows easily that

$$\cosh^2\theta - \sinh^2\theta = 1,$$

and hence $(\cosh\theta, \sinh\theta)$ is a point on the hyperbola $x^2 - y^2 = 1$ for each real value of θ.

By investigating $\cosh\theta$ and $\sinh\theta$ a little more closely, we find that each point on the positive branch ($x > 0$) of $x^2 - y^2 = 1$ occurs as $(\cosh\theta, \sinh\theta)$ for exactly one real value of θ. Some of the relevant properties of the hyperbolic cosine and sine can be seen at a glance from their graphs, shown along with those of $\frac{1}{2}e^\theta$ and $\frac{1}{2}e^{-\theta}$ in Figure 9.8. (Proofs are easily constructed from the fact that e^θ takes each positive value exactly once.)

- $\cosh(-\theta) = \cosh\theta$,
- $\cosh\theta$ takes all real values ≥ 1,
- $\sinh(-\theta) = -\sinh\theta$,
- $\sinh\theta$ takes each real value once.

Thus $\sinh\theta$ takes each real value y exactly once, and for each value $y = \sinh\theta$ the value x of $\cosh\theta$ gives a point (x, y) on the positive branch of $x^2 - y^2 = 1$, because $\cosh^2\theta - \sinh^2\theta = 1$. Because there is exactly one point (x, y) on the positive branch for each y, we therefore get each point on the positive branch exactly once as $(\cosh\theta, \sinh\theta)$.

This one-to-one correspondence between points $P_\theta = (\cosh\theta, \sinh\theta)$ on the positive branch and real numbers θ enables us to "add points" by adding their parameter values. We simply define the *sum of points* $P_\theta = (\cosh\theta, \sinh\theta)$ and $P_\phi = (\cosh\phi, \sinh\phi)$ on $x^2 - y^2 = 1$

$t = \cosh\theta$

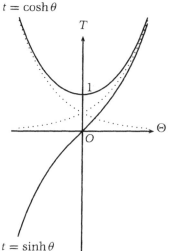

$t = \sinh\theta$

FIGURE 9.8 The graphs of cosh and sinh.

by

$$P_\theta + P_\phi = P_{\theta+\phi} = (\cosh(\theta + \phi), \sinh(\theta + \phi)).$$

Adding points like this may seem an idle and useless thing to do, but we have done it before, with interesting results. The process used in Section 8.5 to generate integer points on $x^2 - 2y^2 = 1$ can be interpreted as repeated "addition" of the point $(3, 2)$ to itself. In the next section we shall see why this is so and how much clearer the process becomes when interpreted as addition of parameter values.

Exercises

The identity $\cosh^2 \theta - \sinh^2 \theta = 1$ is just one of many where the hyperbolic sine and cosine behave almost the same as the ordinary sine and cosine. The similarity is best explained by allowing θ to be complex, so that they all become relatives of the exponential function.

9.5.1. Use Euler's formula for $e^{i\theta}$ and $e^{-i\theta}$ to show that

$$\cos\theta = \frac{e^{i\theta} + e^{-i\theta}}{2} = \cosh i\theta$$

$$\text{and} \quad \sin\theta = \frac{e^{i\theta} - e^{-i\theta}}{2i} = \frac{1}{i}\sinh i\theta.$$

This explains why cosh and sinh satisfy addition formulas, and other identities, similar to those satisfied by cos and sin. Of course, it is not necessary to use complex numbers to *prove* these identities—they follow from properties of the real exponential function—but complex numbers allow us to predict their existence in the first place and to anticipate their form.

9.5.2. Prove the *addition formulas* for cosh and sinh:

$$\cosh(\theta + \phi) = \cosh\theta \cosh\phi + \sinh\theta \sinh\phi,$$

$$\text{and} \quad \sinh(\theta + \phi) = \sinh\theta \cosh\phi + \cosh\theta \sinh\phi.$$

As mentioned earlier, the parameter θ in $x = \cosh\theta$ and $y = \sinh\theta$ can be interpreted as twice the area of a sector of the hyperbola $x^2 - y^2 = 1$. A geometric proof of this fact can be put together as follows. We first

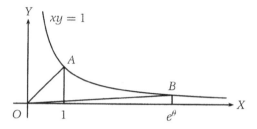

FIGURE 9.9   Areas associated with the hyperbola.

consider the hyperbola $xy = 1$ and the region under it from $x = 1$ to $x = e^\theta$ (Figure 9.9). This region has area θ. (Why?)

9.5.3. Show by geometry that the region bounded by the curve and the lines OA and OB also has area θ.

Now we use the result of exercise 8.3.3 that rotation of $xy = 1$ through angle $\pi/4$ gives the curve $x^2 - y^2 = 2$.

9.5.4. Show that this rotation sends the point $(e^\theta, e^{-\theta})$ on $xy = 1$ to the point $(\sqrt{2}\cosh\theta, -\sqrt{2}\sinh\theta)$ on $x^2 - y^2 = 2$.

9.5.5. By scaling down the curve $x^2 - y^2 = 2$, deduce that the arc of the hyperbola $x^2 - y^2 = 1$ between $(0, 1)$ and $(\cosh\theta, \sinh\theta)$, together with the lines connecting its endpoints to O, bounds a sector of area $\theta/2$.

9.6   The Pell Equation Revisited

The parametric equations $x = \cosh\theta$, $y = \sinh\theta$ for $x^2 - y^2 = 1$ generalize easily to other hyperbolas. The most interesting, from the perspective of this book, are the hyperbolas given by the Pell equation

$$x^2 - dy^2 = 1, \quad \text{where } d \text{ is a nonsquare positive integer.}$$

This equation is satisfied by

$$x = \cosh\theta, \quad y = \frac{1}{\sqrt{d}}\sinh\theta,$$

and in fact the latter equations define a one-to-one correspondence between the real numbers and the points on the positive branch of $x^2 - dy^2 = 1$, as one sees by checking the range and sign of $\cosh\theta$ and $\frac{1}{\sqrt{d}}\sinh\theta$.

Generalizing another idea from the previous section, we let

$$P_\theta = \left(\cosh\theta, \frac{1}{\sqrt{d}}\sinh\theta\right)$$

be the point with parameter value θ, and we define the *sum of points* P_θ and P_ϕ by

$$P_\theta + P_\phi = P_{\theta+\phi}.$$

We can similarly define the difference of points by $P_\theta - P_\phi = P_{\theta-\phi}$.

Now we can work out a rule for computing $P_{\theta+\phi}$ from P_θ and P_ϕ, using the addition formulas for cosh and sinh:

$$\cosh(\theta + \phi) = \cosh\theta \cosh\phi + \sinh\theta \sinh\phi,$$
$$\sinh(\theta + \phi) = \sinh\theta \cosh\phi + \cosh\theta \sinh\phi.$$

As mentioned in the previous exercise set, these addition formulas are very similar to those for cos and sin, but they are easier to prove, because it is only necessary to expand each side in terms of exponentials. It follows from the addition formulas that

$$P_{\theta+\phi} = \left(\cosh(\theta + \phi), \frac{1}{\sqrt{d}}\sinh(\theta + \phi)\right)$$

$$= \Big(\cosh\theta \cosh\phi + \sinh\theta \sinh\phi,$$

$$\frac{1}{\sqrt{d}}(\sinh\theta \cosh\phi + \cosh\theta \sinh\phi)\Big).$$

This gives a rule to compute the coordinates of $P_{\theta+\phi}$ from the coordinates of P_θ and P_ϕ. The rule can be stated concisely as follows.

Rule for adding points on $x^2 - dy^2 = 1$.
If

$$P_\theta = \left(\cosh\theta, \frac{1}{\sqrt{d}}\sinh\theta\right) = (x_\theta, y_\theta)$$

and

$$P_\phi = \left(\cosh \phi, \frac{1}{\sqrt{d}} \sinh \phi\right) = (x_\phi, y_\phi),$$

then

$$P_{\theta+\phi} = (x_{\theta+\phi}, y_{\theta+\phi}),$$

where

$$x_{\theta+\phi} = x_\theta x_\phi + dy_\theta y_\phi \quad and \quad y_{\theta+\phi} = x_\theta y_\phi + y_\theta x_\phi. \qquad \square$$

In the special case where $P_\theta = (x_\theta, y_\theta)$ and $P_\phi = (x_\phi, y_\phi)$ are integer points, we notice that $P_{\theta+\phi}$ is also an integer point, and its coordinates $x_{\theta+\phi}$ and $y_{\theta+\phi}$ are the integers defined by

$$x_{\theta+\phi} + y_{\theta+\phi}\sqrt{d} = (x_\theta + y_\theta\sqrt{d})(x_\phi + y_\phi\sqrt{d}).$$

This follows immediately by expanding the right-hand side to

$$x_\theta x_\phi + dy_\theta y_\phi + (x_\theta y_\phi + y_\theta x_\phi)\sqrt{d}$$

and comparing with the rule for adding points.

Thus *the rule previously used to generate integer points on $x^2 - dy^2 = 1$ (Section 8.5) is the same as the rule for adding points by their parameter values.*

This wonderful pre-established harmony between arithmetic and geometry gives a new and useful view of the integer points on the positive branch of $x^2 - dy^2 = 1$. It shows that they form a *subgroup* of the abelian group of all points on the positive branch. The points P_θ on the positive branch are an abelian group because of the way they correspond to real numbers θ: they inherit their $+$ operation from ordinary $+$ on $\mathbb{R}$, and with it the abelian group properties of ordinary $+$ we observed in Section 6.10. And the integer points on the positive branch are also a group because:

- the sum of integer points is an integer point, by the rule for adding points,
- the inverse of an integer point (x, y) is an integer point, because

$$(x + y\sqrt{d})^{-1} = \frac{x - y\sqrt{d}}{x^2 - dy^2} = x - y\sqrt{d},$$

because $x^2 - dy^2 = 1$ by hypothesis.

In fact, the integer points form an *infinite cyclic* group   by the following result.

Generation of integer points on $x^2 - dy^2 = 1$**.**   *The integer points on* $x^2 - dy^2 = 1$ *all result from* $(1, 0)$ *and the integer point nearest to it by addition and subtraction of points.*

Proof   Consider the group of integer points P_ϕ on $x^2 - dy^2 = 1$, and the corresponding group of real numbers ϕ. It will suffice to show that the latter group consists of the integer multiples of θ, because the multiples of θ are precisely the numbers resulting from θ and 0 by addition and subtraction, and hence the corresponding points are those resulting from P_θ and $P_0 = (1, 0)$ by addition and subtraction of points.

   The crucial property of the group of integer points P_ϕ is that it has a member nearest to $(1, 0)$, because integer points cannot approach arbitrarily close to $(1, 0)$, and hence the corresponding group of real numbers has a member θ closest to 0. Such a group of reals consists of the integer multiples $n\theta$ of θ. Why? Because if ϕ were a member strictly between, say, $k\theta$ and $(k+1)\theta$ then $\phi - k\theta$ would be a member of the group strictly between 0 and θ, contrary to the choice of θ.   □

   This theorem implies that the solutions of $x^2 - 2y^2 = 1$ we found in Section 8.5, by adding the point $(3, 2)$ to itself, are in fact *all* the integer solutions with x and y positive, because $(3, 2)$ is the nearest integer point to $(1, 0)$. Similarly, all the solutions of $x^2 - 3y^2 = 1$ with x and y positive are found by adding the point $(2, 1)$ to itself. A similar result holds on $x^2 - dy^2 = 1$ for any nonsquare positive integer d, thanks to the result of Section 8.7*.

Exercises

The idea of this section can be extended to other hyperbolas, even those that result from rotation of axes. An interesting example is the hyperbola $x^2 + xy - y^2 = 1$, which Vajda (1989) p. 34 showed to contain the integer points (F_{2n-1}, F_{2n}), where the F_k are the *Fibonacci numbers* defined by

$$F_0 = 0, \quad F_1 = 1, \quad F_{k+1} = F_k + F_{k-1}.$$

These numbers are linked in many ways to the roots $\tau = (1 + \sqrt{5})/2$ and $\tau^* = (1 - \sqrt{5})/2$ of the equation $t^2 = t + 1$, and so is the curve $x^2 + xy - y^2 = 1$.

9.6.1. Show that $x^2 + xy - y^2 = (x + y\tau)(x + y\tau^*)$.

The irrational number $1 + \tau$ can be used to generate the pairs (F_{2n-1}, F_{2n}) of consecutive Fibonacci numbers, much as $1 + \sqrt{2}$ was used to generate side and diagonal numbers in Exercise 8.5.3.

9.6.2. Deduce from the definition of Fibonacci numbers that $F_{k+3} = 2F_{k+1} + F_k$. Use this formula to prove by induction that $(1 + \tau)^n = F_{2n-1} + F_{2n}\tau$ for all natural numbers n.

Now we take the hint from the factorization in Exercise 9.6.1 by defining the *sum of integer points* (x_1, y_1) *and* (x_2, y_2) *on* $x^2 + xy - y^2 = 1$ to be (x_3, y_3), where x_3 and y_3 are the integers satisfying

$$x_3 + y_3\tau = (x_1 + y_1\tau)(x_2 + y_2\tau).$$

9.6.3. Check that if (x_1, y_1) and (x_2, y_2) are points on $x^2 + xy - y^2 = 1$ then so is their sum (x_3, y_3).

9.6.4. Use addition of points to show that the hyperbola $x^2 + xy - y^2 = 1$ contains all the points (F_{2n-1}, F_{2n}) for positive integers n.

The equation $x^2 + xy - y^2 = 1$ can be rewritten $\left(x + \frac{y}{2}\right)^2 - \left(\frac{\sqrt{5}}{2}y\right)^2 = 1$ by completing the square, which suggests the parametric equations

$$x + \frac{y}{2} = \cosh\theta, \quad \frac{\sqrt{5}}{2}y = \sinh\theta.$$

9.6.5. Deduce from these parametric equations that $x + y\tau = e^\theta$, and hence conclude that the rule for adding points on $x^2 + xy - y^2 = 1$ amounts to adding their parameter values θ.

But adding parameter values makes sense for *any* points on the parameterized branch of the hyperbola, so we can extend addition of integer points to addition of any points on this branch.

As with the Pell equation, this extended addition operation allows us to find all the integer points on the branch. First we throw in some new ones—the inverses (with respect to addition of points) of the points previously found.

9.6.6. Show that $(1 + \tau)^{-1} = 2 - \tau$, and hence show by induction that $(1 + \tau)^{-n} = F_{2n+1} - F_{2n}\tau$ for all positive integers n.

9.6.7. Deduce from Exercises 9.6.6 and 9.6.3 that the points $(F_{2n+1}, -F_{2n})$ are on the curve $x^2 + xy - y^2 = 1$ for all positive integers n.

Finally, we see that the integer points found so far are the only ones on the branch, by relating them to a subgroup of the real numbers.

9.6.8. Show that the integer points (x_n, y_n) on $x^2 + xy - y^2 = 1$ defined by $x_n + y_n\tau = (1 + \tau)^n$ for all integers n form a subgroup of all the points on $x^2 + xy - y^2 = 1$ under addition of points, and deduce that they are *all* the integer points on the branch containing them.

9.7 Discussion

There is no last word on numbers and geometry, because these themes have infinite depth and variety. Nevertheless, this last chapter is a high point of sorts, from which it is possible to survey the ideas we have developed so far and give them some order and direction. I shall therefore discuss the ideas of Chapter 9 against the background of the whole book, reviewing some trains of thought that have led us to this point and suggesting how they might be pursued further.

From Natural Numbers to Complex Numbers

The long march of the number concept from $\mathbb{N}$ to $\mathbb{C}$, from counting to geometry, is one of the great sagas of mathematics. It was a struggle against almost insuperable obstacles, and every learner relives the struggle, to some extent, even when following the marked trail. Unfortunately, those already in the know tend to forget this. Once overcome, an obstacle is no longer an obstacle, only a step up to the next level. But making sense of negative, irrational, and imaginary numbers once seemed *impossible* tasks, so it is worth reflecting on the power of the ideas that made them possible.

The step from $\mathbb{N}$ to the integers $\mathbb{Z}$ or the rationals $\mathbb{Q}$ is technically a small one today, now that we are happy to accept infinite sets as mathematical objects. An integer can be defined as a set of pairs of natural numbers with constant difference, for example,

$$+3 = \{(0, 3), (1, 4), (2, 5), (3, 6), \ldots\}$$
$$-3 = \{(3, 0), (4, 1), (5, 2), (6, 3), \ldots\}.$$

We can similarly use sets of pairs of integers with constant quotient to define rational numbers. The real point of expanding $\mathbb{N}$ to $\mathbb{Z}$ and $\mathbb{Q}$ is the simplification in structure, allowing unlimited subtraction and division (except division by 0). The structure obtained—a ring in the case of $\mathbb{Z}$, a field in the case of $\mathbb{Q}$—turns out to be one of the most fruitful ideas in mathematics. It allows our intuition about numbers to be used in the study of congruence classes, polynomials, and even more abstract objects. The field structure also guides further extensions of the number concept; it becomes precisely the thing we want to preserve.

The step from $\mathbb{Q}$ to $\mathbb{R}$ is the most profound, because it creates a model of the real line and hence bridges the gap between discrete and continuous. It is this step that commits us to set theory irrevocably; we cannot do without infinite sets and an uncountable number of them. However, this step also brings a huge gain in understanding. It not only shows that rational numbers are "rare," in the sense that there are only countably many of them, it shows that the same is true of the *algebraic* numbers; the numbers (such as $\sqrt{2}$) that are roots of polynomial equations with integer coefficients.

When Cantor proved that $\mathbb{R}$ is uncountable in 1874 he actually began by proving that there are only countably many algebraic numbers (which is not hard, because it amounts to listing the polynomial equations with integer coefficients). His uncountability proof then enabled him to conclude immediately that *transcendental numbers exist*, a result previously obtained with great difficulty by proving specific numbers transcendental.

Dedekind's idea of creating $\mathbb{R}$ by completing $\mathbb{Q}$, or "filling its gaps," also plugged many holes in the logic of calculus and geometry. In particular, it is crucial in the so-called *fundamental theorem of algebra*, which states that any polynomial equation has a root in $\mathbb{C}$. This turns out to depend on properties of continuous functions on the plane, which were not properly proved until the completeness of $\mathbb{R}$ was understood. The details may be found for example in Stillwell (1994).

Extending $\mathbb{R}$ to $\mathbb{C}$ is technically a small step. As mentioned in Section 7.1, it suffices to define complex numbers as pairs of real numbers and define $+$ and $\times$ appropriately. The unexpected power of $\mathbb{C}$ derives partly from the completeness of $\mathbb{R}$ and partly from the harmony between $+$ and $\times$ on $\mathbb{C}$ and the geometry of the plane. It is a kind of miracle that so much algebraic and geometric structure can coexist in the same object, and in fact this miracle is not repeated. There is no way to define $+$ and $\times$ in a space of three or more dimensions so that the resulting structure is a field, though there are "near misses" in dimension 4 (the multiplication is noncommutative) and dimension 8 (the multiplication is nonassociative and noncommutative). More information on this interesting question may be found in Artmann (1988) and Ebbinghaus et al. (1991).

The Exponential Function

The exponential function, like $\mathbb{C}$ itself, is an amazing confluence of arithmetic and geometric ideas.

In arithmetic, the idea of exponentiation first arises in $\mathbb{N}$. As we know from Section 1.7, powers of 2 occur in Euclid's theorem on perfect numbers. The fundamental property of powers of 2 is that when powers are multiplied, their exponents add:

$$2^m \times 2^n = 2^{m+n}.$$

In Section 6.10 we observed that in $\mathbb{Q}$, where we also have negative powers of 2, there is in fact an *isomorphism* between the group of powers of 2 under $\times$ and the group of integers under $+$. One of the advantages of $\mathbb{R}$ is that it allows 2^a to be defined for any real numbers a, extending the exponent addition property to all real a and b:

$$2^a \times 2^b = 2^{a+b}.$$

This gives an isomorphism between the group of real powers of 2, under $\times$, and the group $\mathbb{R}$ under $+$.

This isomorphism allows us to multiply positive reals x and y by the simpler operation of addition as soon as we know *logarithms* and *antilogarithms*. The logarithm (to base 2) of $x > 0$ is the number a such that $x = 2^a$, so if we also know the logarithm b of y we can find

$x \times y$ from

$$x \times y = 2^a \times 2^b = 2^{a+b}$$

by forming the sum $a+b$ and looking up its "antilogarithm" 2^{a+b}. The logarithm function was originally invented for this purpose, and only later found to be the area under the hyperbola. It seems even more remarkable that the complex exponential function $e^{i\theta}$ turned out to be $\cos\theta + i\sin\theta$ (Euler (1748)). However, hints of this relationship had been around for centuries, and we have seen some of them, for example, Viète's "product of triangles" discussed in Section 7.2.

Perhaps the most remarkable thing is that e^z can be defined, for complex z, in a completely geometric manner. Consider the problem of mapping the *cartesian coordinate grid*, of lines $x = $ constant and $y = $ constant, onto the *polar coordinate grid* of radial lines $\theta = $ constant and concentric circles $r = $ constant. Putting $z = x + iy$, we see that $e^z = e^x(\cos y + i\sin y)$ does the trick: the line $x = $ constant is mapped onto the circle of radius e^x, and the line $y = $ constant is mapped onto the ray $\theta = y$.

This map *preserves angles*, not through any special merit of e^z, because in fact the same is true of most of the complex functions one meets. In this case, one can see immediately that the right angles between the lines $x = $ constant and $y = $ constant map to right angles between the radial lines and the circles in the polar coordinate grid.

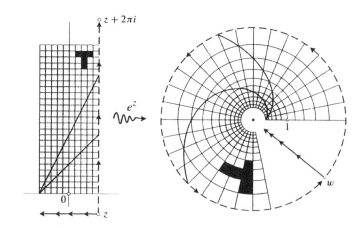

FIGURE 9.10 The exponential map.

It is more interesting to observe what happens to a diagonal through the "little squares" in the cartesian coordinate grid (Figure 9.10). It is mapped to an *equiangular spiral* through the "little quadrilaterals" in the polar coordinate grid, and these "quadrilaterals" have to grow exponentially in size to maintain the constant angle of the spiral.

Conversely, any angle-preserving map from the cartesian grid to the polar grid forces exponential growth on the image circles $r =$ constant. Thus if one starts with the circle and the concept of angle, one is led inexorably to the arithmetic of exponentiation. More about the relation between geometric and arithmetic properties of complex functions may be found in the beautiful book of Needham (1997), from which Figure 9.10 is taken.

From Pythagoras to Pell

The two most important quadratic equations in this book are the Pythagorean equation

$$a^2 + b^2 = c^2$$

and the Pell equation

$$x^2 - dy^2 = 1.$$

We have seen how Pythagoras' theorem leads to the discovery that the diagonal of the unit square is $\sqrt{2}$, which in turn confronts us with the problem of understanding the irrational number $\sqrt{2}$.

This leads to the Pell equation $x^2 - 2y^2 = 1$, whose integer solutions $x = x_n$, $y = y_n$ give a sequence of rationals x_n/y_n converging to $\sqrt{2}$. The same rationals arise from the continued fraction

$$\sqrt{2} = 1 + \cfrac{1}{2 + \cfrac{1}{2 + \cfrac{1}{2 + \cfrac{1}{\ddots}}}}.$$

and the latter is surely the simplest "explanation" of $\sqrt{2}$ in terms of rational numbers.

In attempting to understand $\sqrt{d}$, for any nonsquare positive integer d, we follow a similar approach: find the solutions of the Pell equation $x^2 - dy^2 = 1$, and the continued fraction for $\sqrt{d}$. The emphasis shifts, however, as one discovers that understanding is helped by actual use of the number $\sqrt{d}$. In fact, it is helpful to use all the real numbers and the transcendental functions cosh and sinh. This is the message of Section 9.6.

There we gave an arbitrary point (x, y) on the hyperbola $x^2 - dy^2 = 1$ the coordinate θ such that

$$x = \cosh\theta, \quad y = \frac{1}{\sqrt{d}}\sinh\theta.$$

We represented each *integer* point (x, y) by the quadratic integer $x + y\sqrt{d}$. Notice that

$$x + y\sqrt{d} = \cosh\theta + \sinh\theta = e^\theta,$$

so the integer points are represented by quadratic integer values of the exponential function e^θ.

The quadratic integers $x + y\sqrt{d}$ such that $x^2 - dy^2 = 1$ have the property that the product of any two of them, $x_1 + y_1\sqrt{d}$, $x_2 + y_2\sqrt{d}$, is another quadratic integer $x_3 + y_3\sqrt{d}$ with the same property. It follows that they form an abelian group under $\times$, and their logarithms form an abelian group under $+$. The essence of the proof in Section 9.6 is to use logarithms to convert the group of quadratic integers with the $\times$ operation to the group of θ values with the $+$ operation, which is easier to understand.

In this instance it is possible to understand the multiplicative group without the help of logarithms, and many number theory books do this. However, the current proof has certain merits. The exponential function is naturally associated with the hyperbola $x^2 - dy^2 = 1$ anyway, and the proof is a model for a more general theorem in which logarithms are always used: *Dirichlet's unit theorem*. The Pell equation solution is essentially the one-dimensional case of this theorem; the general case may be found in books on algebraic number theory, for example, Samuel (1970).

The solution of $x^2 - dy^2 = 1$ shows that all solutions are generated in a simple way from the smallest nontrivial solution, but the nature of this smallest solution remains a mystery. Dirichlet's pigeonhole argument shows that it must exist (Section 8.7*) but does not relate it to d in any reasonable way. In fact, its dependence on d is highly irregular, judging from notorious values like $d = 61$, for which the smallest positive solution is $x = 1766319049$, $y = 226153980$.

There is probably no *simple* relationship between d and the smallest solution, but there is an extremely interesting relationship, also discovered by Dirichlet. It is called his *class number formula*, and it relates the smallest solution of $x^2 - dy^2 = 1$ to the so-called *class number* of the quadratic integers $x + y\sqrt{d}$, a measure of their deviation from unique prime factorization. For an introduction to this deep and complicated subject, see Scharlau and Opolka (1985), or see Dirichlet's own treatment in Dirichlet (1867).

The hidden depths of the Pell equation opened into a yawning chasm in recent decades, with unexpected discoveries in mathematical logic. Since the time of Lagrange, mathematicians have known general algorithms for finding integer solutions of quadratic equations, or more importantly, deciding whether solutions exist. (The solution of the Pythagorean equation was the first success in this field, and Lagrange found that the solution of the Pell equation opened the door to all other quadratics.) But no general algorithms for higher-degree equations were ever discovered, and in 1970 Yuri Matijasevič proved that *there is no such algorithm*. The idea of his proof is to show that polynomial equations are complex enough to "simulate" arbitrary computations, because results from logic tell us that no algorithm can answer all questions about computation. The biggest technical difficulty is finding a single equation that is manageable yet sufficiently complex. It turns out to be none other than the Pell equation! The proof, which has now been boiled down to simple number theory, may be seen in Jones and Matijasevič (1991).

Bibliography

Artin, M. (1991). *Algebra*. Prentice-Hall.

Artmann, B. (1988). *The Concept of Number*. Halsted Press.

Baker, A. (1984). *A Concise Introduction to the Theory of Numbers*. Cambridge University Press.

Beltrami, E. (1868). Teoria fondamentale degli spazii di curvatura costante. *Annali di Matematica pura et applicata, series II*, **II**, 232–255. English translation, *Fundamental Theory of Spaces of Constant Curvature*, in Stillwell (1996), 41–62.

Bombelli, R. (1572). *L'Algebra*. Reprinted by Feltrinelli Editore, Milan, 1966.

Cassels, J. W. S. (1991). *Lectures on Elliptic Curves*. Cambridge University Press.

Colebrooke, H. T. (1817). *Algebra, with Arithmetic and Mensuration, from the Sanscrit of Brahmegupta and Bháskara*. John Murray, London. Reprinted by Martin Sandig, Wiesbaden, 1973.

Cormen, T. H., Leiserson, C. E., and Rivest, R. L. (1990). *Introduction to Algorithms*. MIT Press.

Dedekind, R. (1872). *Stetigkeit und Irrationalzahlen*. English translation, *Continuity and Irrational Numbers*, in *Essays on the Theory of Numbers*, Open Court, 1901.

Dedekind, R. (1877). *Sur la Théorie des Nombres Entiers Algébriques*. Gauthier-Villars. English translation, *Theory of Algebraic Integers*, Cambridge University Press, 1996.

Dedekind, R. (1888). *Was sind und was sollen die Zahlen?* Braunschweig. English translation in *Essays on the Theory of Numbers*, Open Court, Chicago, 1901.

Descartes, R. (1637). *La Géométrie*. English translation, *The Geometry*, Dover, 1954.

Dirichlet, P. G. L. (1837). Beweis der Satz, dass jede unbegrentze arithmetische Progression, deren erstes Glied und Differenz ganze Zahlen ohne gemeinschaftliche Factor sind, unendliche viele Primzahlen enthält. *Abhandlungen der Königliche Preussische Akademie der Wissenschaften*, pages 45–81. Also in his *Mathematische Werke*, volume 1, 313–342.

Dirichlet, P. G. L. (1867). *Vorlesungen über Zahlentheorie*. Vieweg. Fourth edition (1894) reprinted by Chelsea, New York, 1968.

Ebbinghaus, H.-D., Hermes, H., Hirzebruch, F., Koecher, M., Mainzer, K., Neukirch, J., Prestel, A., and Remmert, R. (1991). *Numbers*. Springer-Verlag.

Euler, L. (1748). *Introductio in analysin infinitorum*. In his *Opera Omnia*, series 1, volume VIII. English translation, *Introduction to the Analysis of the Infinite*, Springer-Verlag, 1988.

Euler, L. (1770). *Vollständige Einleitung zur Algebra*. In his *Opera Omnia*, series 1, volume I. English translation, *Elements of Algebra*, Springer–Verlag, 1984.

Gauss, C. F. (1801). *Disquisitiones Arithmeticae*. In his *Werke*, volume 1. English edition published by Springer-Verlag, 1986.

Grassmann, H. (1861). Stücke aus dem Lehrbuche der Arithmetik. *Hermann Grassmann's Mathematische und Physkalische Werke*, **II/1**, 295–349.

Guy, R. K. (1994). *Unsolved Problems in Number Theory*. Springer-Verlag.

Hasse, H. (1928). Über eindeutige Zerlegung in Primelemente oder in Primhauptideal in Integrätsbereichen. *Journal für reine und angewandte Mathematik*, **159**, 3–12.

Heath, T. L. (1910). *Diophantus of Alexandria*. Cambridge University Press. Reprinted by Dover, 1964.

Heath, T. L. (1912). *The Works of Archimedes*. Cambridge University Press. Reprinted by Dover, 1953.

Heath, T. L. (1925). *The Thirteen Books of Euclid's Elements*. Cambridge University Press. Reprinted by Dover, 1956.

Hilbert, D. (1899). *Grundlagen der Geometrie*. B. G. Teubner, Stuttgart. English translation, *Foundations of Geometry*, Open Court, Chicago, 1971.

Jones, G. A., and Singerman, D. (1987). *Complex Functions*. Cambridge University Press.

Jones, J. P., and Matijasevič, Y. V. (1991). Proof of the recursive unsolvability of Hilbert's tenth problem. *American Mathematical Monthly*, **98**, 689–709.

Katz, V. (1993). *A History of Mathematics*. HarperCollins.

Koblitz, N. (1985). *Introduction to Elliptic Curves and Modular Forms*. Springer-Verlag.

Magnus, W. (1974). *Noneuclidean Tesselations and Their Groups*. Academic Press.

Needham, T. (1997). *Visual Complex Analysis*. Oxford University Press.

Neugebauer, O., and Sachs, A. (1945). *Mathematical Cuneiform Texts*. Yale University Press.

Niven, I., Zuckerman, H. S., and Montgomery, H. L. (1991). *An Introduction to the Theory of Numbers*. Wiley.

Pascal, B. (1654). Traité du triangle arithmétique. *Pascal Œuvres Complètes*. Éditions du Seuil, 1963. English translation, Treatise on the Arithmetic Triangle, in *Great Books of the Western World: Pascal*, Encyclopedia Britannica, 1952, 447–473.

Poincaré, H. (1882). Théorie des groupes fuchsiens. *Acta Mathematica*, **1**, 1–62. An English translation of the relevant part is in Stillwell (1996), 123–129.

Poincaré, H. (1883). Mémoire sur les groupes kleinéens. *Acta Mathematica*, **3**, 49–92. An English translation of the relevant part is in Stillwell (1996), 131–137.

Rousseau, G. (1991). On the quadratic reciprocity law. *Journal of the Australian Mathematical Society*, **51**, 423–425.

Samuel, P. (1970). *Algebraic Theory of Numbers*. Hermann.

Scharlau, W., and Opolka, H. (1985). *From Fermat to Minkowski: Lectures on the Theory of Numbers and Its Historical Development*. Springer-Verlag.

Shallit, J. (1994). Origins of the analysis of the Euclidean algorithm. *Historia Mathematica*, **21**, 401–419.

Sigler, L. E. (1987). *Leonardo Pisano Fibonacci. The Book of Squares*. Academic Press. English translation of Fibonacci's *Liber quadratorum*.

Stark, H. M. (1978). *An Introduction to Number Theory*. MIT Press.

Stillwell, J. C. (1992). *Geometry of Surfaces*. Springer-Verlag.

Stillwell, J. C. (1994). *Elements of Algebra*. Springer-Verlag.

Stillwell, J. C. (1996). *Sources of Hyperbolic Geometry*. American Mathematical Society.

Taylor, A. E. (1972). *Plato. Philebus and Epinomis*. Thomas Nelson and Sons Ltd.

Vajda, A. (1989). *Fibonacci & Lucas Numbers and the Golden Section*. Halsted Press.

Viète, F. (1579). *Universalium inspectionium ad canonem mathematicum liber singularis*.

von Neumann, J. (1923). Zur Einführung der transfiniten Zahlen. *Acta litterarum ac scientiarum Regiae Universitas Hungaricae Francisco-Josephine, Sectio scientiarum mathematicarum*, pages 199–208. English translation in *From Frege to Gödel* (Ed. J. van Heijenoort), Harvard University Press, 1967, 347–354.

Wallis, J. (1655). *Arithmetica infinitorum*. In his *Opera*, volume 1, 355–478.

Weil, A. (1984). *Number Theory. An Approach through History*. Birkhäuser.

Young, R. M. (1992). *Excursions in Calculus*. Mathematical Association of America.

Index

317

Undergraduate Texts in Mathematics

(continued after index)

Undergraduate Texts in Mathematics

(continued from page ii)

Hilton/Holton/Pedersen: Mathematical Reflections: In a Room with Many Mirrors.

Iooss/Joseph: Elementary Stability and Bifurcation Theory. Second edition.

Isaac: The Pleasures of Probability. *Readings in Mathematics.*

James: Topological and Uniform Spaces.

Jänich: Linear Algebra.

Jänich: Topology.

Kemeny/Snell: Finite Markov Chains.

Kinsey: Topology of Surfaces.

Klambauer: Aspects of Calculus.

Lang: A First Course in Calculus. Fifth edition.

Lang: Calculus of Several Variables. Third edition.

Lang: Introduction to Linear Algebra. Second edition.

Lang: Linear Algebra. Third edition.

Lang: Undergraduate Algebra. Second edition.

Lang: Undergraduate Analysis.

Lax/Burstein/Lax: Calculus with Applications and Computing. Volume 1.

LeCuyer: College Mathematics with APL.

Lidl/Pilz: Applied Abstract Algebra. Second edition.

Logan: Applied Partial Differential Equations.

Macki-Strauss: Introduction to Optimal Control Theory.

Malitz: Introduction to Mathematical Logic.

Marsden/Weinstein: Calculus I, II, III. Second edition.

Martin: The Foundations of Geometry and the Non-Euclidean Plane.

Martin: Geometric Constructions.

Martin: Transformation Geometry: An Introduction to Symmetry.

Millman/Parker: Geometry: A Metric Approach with Models. Second edition.

Moschovakis: Notes on Set Theory.

Owen: A First Course in the Mathematical Foundations of Thermodynamics.

Palka: An Introduction to Complex Function Theory.

Pedrick: A First Course in Analysis.

Peressini/Sullivan/Uhl: The Mathematics of Nonlinear Programming.

Prenowitz/Jantosciak: Join Geometries.

Priestley: Calculus: A Liberal Art. Second edition.

Protter/Morrey: A First Course in Real Analysis. Second edition.

Protter/Morrey: Intermediate Calculus. Second edition.

Roman: An Introduction to Coding and Information Theory.

Ross: Elementary Analysis: The Theory of Calculus.

Samuel: Projective Geometry. *Readings in Mathematics.*

Scharlau/Opolka: From Fermat to Minkowski.

Schiff: The Laplace Transform: Theory and Applications.

Sethuraman: Rings, Fields, and Vector Spaces: An Approach to Geometric Constructability.

Sigler: Algebra.

Silverman/Tate: Rational Points on Elliptic Curves.

Simmonds: A Brief on Tensor Analysis. Second edition.

Singer: Geometry: Plane and Fancy.

Singer/Thorpe: Lecture Notes on Elementary Topology and Geometry.

Smith: Linear Algebra. Third edition.

Smith: Primer of Modern Analysis. Second edition.

Stanton/White: Constructive Combinatorics.

Stillwell: Elements of Algebra: Geometry, Numbers, Equations.

Stillwell: Mathematics and Its History.

Stillwell: Numbers and Geometry. *Readings in Mathematics.*

Strayer: Linear Programming and Its Applications.

Thorpe: Elementary Topics in Differential Geometry.

Toth: Glimpses of Algebra and Geometry. *Readings in Mathematics.*

Troutman: Variational Calculus and Optimal Control. Second edition.

Valenza: Linear Algebra: An Introduction to Abstract Mathematics.

Whyburn/Duda: Dynamic Topology.

Wilson: Much Ado About Calculus.

CPSIA information can be obtained at www.ICGtesting.com
Printed in the USA
LVOW011014151212

311808LV00007B/207/A